T0073222

TIME

AND

BEAUTY

Why Time Flies and Beauty Never Dies

Related Books by Adrian Bejan

Freedom and Evolution: Hierarchy in Nature, Society and Science
Springer Nature, 2020
ISBN: 978-3030340087

The Physics of Life: The Evolution of Everything
St. Martin's Press, 2016
ISBN: 978-1250078827

Design in Nature: How the Constructal Law Governs Evolution in Biology, Physics, Technology, and Social Organization
with J. P. Zane, Doubleday, 2012
ISBN: 978-0307744340

Advanced Engineering Thermodynamics, 4th Edition
Wiley, 2016
ISBN: 978-1-119-05209-8

Convection Heat Transfer, 4th Edition
Wiley, 2013
ISBN: 978-0-470-90037-6

Design with Constructal Theory
with S. Lorente, Wiley, 2008
ISBN: 978-0-471-99816-7

Shape and Structure, from Engineering to Nature
Cambridge University Press, 2000
ISBN: 978-0-521-79388-9

TIME
AND
BEAUTY

Why Time Flies and Beauty Never Dies

ADRIAN BEJAN

Duke University, USA

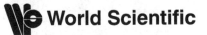

NEW JERSEY · LONDON · SINGAPORE · BEIJING · SHANGHAI · HONG KONG · TAIPEI · CHENNAI · TOKYO

Published by

World Scientific Publishing Co. Pte. Ltd.

5 Toh Tuck Link, Singapore 596224

USA office: 27 Warren Street, Suite 401-402, Hackensack, NJ 07601

UK office: 57 Shelton Street, Covent Garden, London WC2H 9HE

Library of Congress Cataloging-in-Publication Data

Names: Bejan, Adrian, 1948–   author.

Title: Time and beauty : why time flies and beauty never dies / Adrian Bejan.

Description: New Jersey : World Scientific, 2022. | Includes bibliographical references.

Identifiers: LCCN 2021050508 | ISBN 9789811245466 (hardcover) |
    ISBN 9789811246791 (paperback) | ISBN 9789811245473 (ebook for institutions) |
    ISBN 9789811245480 (ebook for individuals)

Subjects: LCSH: Time perception. | Aesthetics--Physiological aspects. | Form perception. |
    Perception--Physiological aspects. | Cognition--Age factors.

Classification: LCC QP445 .B45 2022 | DDC 153.7/53--dc23/eng/20211207

LC record available at https://lccn.loc.gov/2021050508

British Library Cataloguing-in-Publication Data

A catalogue record for this book is available from the British Library.

For any available supplementary material, please visit
https://www.worldscientific.com/worldscibooks/10.1142/12506#t=suppl

Typeset by Stallion Press

Email: enquiries@stallionpress.com

Acknowledgement

I thank my wife Mary and our Cristina, Teresa and William, for the love and encouragement to write this book. I also thank Deborah Fraze for assisting me in my work and for typing the manuscript. The Lane Family Ethics in Technology program supported the writing of Chapter 6 of this book.

I am grateful to Profs. Umit Gunes, Abdulrahman Almerbati, Sinan Gucluer, Hamad Almahmoud, Marcelo Errera, Sylvie Lorente, George Tsatsaronis, Mohamed Awad, Cesare Biserni, Pezhman Mardanpour, Tanmay Basak, Ahmed Waheed and Erdal Cetkin for their concrete support during the past two years.

I am fortunate to be close to free thinkers who take an interest in my ideas, and teach me how to think, speak and write better: Peder Zane, Christopher Davis, Soh Yong Qi, Victor Niederhoffer, Deborah Patton, Matthew Futterman, Ian Coville, Simeon Mihaylov, Malcolm Dean, and David Troy.

In this book on beauty and science, I was inspired by the artists who have created the "constructal art" current: Juana Gomez, Maria Santos Blanco, Gabriela Torres, Star Bales, Luisa Cervantes, Christine Forni, Kris Vanston and Fred Westbrook.

Contents

1
Overview

Growing up, I realized that time was passing faster and faster. This feeling persisted. I became so curious that in March 2019 I published a physics-based explanation for why this feeling is so common. My article [1] was the most read in the entire world during that month [2]. The obvious message is that aging is of concern to all of us on the planet. This book is my story of the idea and its reflection in the physics of beauty and usefulness in life.

Time is slipping through our fingers and never comes back. It slips faster as we get older. Have you ever wondered why we feel this way? Have you questioned other perceptions? For example, have you noticed that most of us are attracted to images that are roughly 50 percent longer horizontally than vertically, as shown in Figure 1.1? Beautiful paintings in art galleries tend to be shaped this way [3]. Have you ever wondered why we are attracted to beauty?

Why these common observations? I questioned them, and I want to share with you the thread, the one idea that connects them all. As you begin to see this connection, you will acquire a deeper understanding of the biggest and most divisive enigma in science today: evolution.

Time and beauty are two of our most basic perceptions. They are so common that they are overlooked. Obviously, they

Figure 1.1   Time slips through our fingers, and beauty is in the horizontal, golden ratio shape of the image (photo: Ben White, @benwhite photograph).

are poles apart. Time is not to be confused with beauty. To paint time is as absurd as to clock beauty.

Different does not mean unrelated.

In this book I show that these two perceptions — the feeling that time accelerates with age, and the attraction of beauty — are both part of the same natural design, in accord with the constructal law[1]. They are an integral aspect of the human animal, striving to live and improve along the way. The perceptions of time and beauty owe their origin to the physical configuration of the currents that convey observations to the brain, and to the human tendency to make choices that empower us to live more and more easily.

Time and beauty are physics. In this book, many aspects of the connection between time and beauty are organized under three principal ideas:

---

[1] For a finite size flow system to persist in time (to live) it must *evolve* with *freedom* such that it provides easier and greater access to what flows (to read more, see the books on page 186).

The first is our perception of time and why we feel that time flies faster as we get older. The perceived time is called "mind time," and it is different from the clock time. The mind time belongs to the individual, the observer, and is a sequence of images — reflections of nature — that are fed by stimuli from the sensory organs.

The rate at which changes in mental images are perceived decreases with age because of several physical features that change with age: the saccade frequency decreases, the body size increases, the pathway degradation intensifies, and so on. The misalignment between mental image time and clock time serves to unify the voluminous body of published observations on evolution as a phenomenon and principle of physics.

Second, beauty attracts because beautifully shaped images are scanned faster by the two eyes. To observe the immediate surroundings and understand them faster is good for life. The book illustrates the main features of image creation and transmission, such as shape, contrast, message, and perspective.

Third, the book puts time and beauty together in order to explain why at the start of the pandemic it felt as if the mind time slowed down. The explanation leads to techniques that each of us can use to slow down our accelerating mind time, and to achieve the feeling that we live longer and more creatively.

Here is a brief run-through of the chapters in the book:

Observations that make us conscious of the passage of time are limitless (Chapter 2). Change is omnipresent, and therefore easy to notice. In fact, animal design is about perceiving the changes in the immediate surroundings that constitute the animal's "niche". We breathe at particular time intervals between inhalation and exhalation. We can vary our breathing times if instructed to do so during a pulmonary exam, yet it is natural to breathe at certain intervals that are so natural that we do not think we are performing work as we breathe.

The same holds for blood circulation. Our hearts beat at regular time intervals. We cannot control that frequency, yet it varies depending on what the rest of the body and its niche do. The athlete's heart beats faster during the game than on the side lines or in bed at night.

Time and timing are everywhere. We see this in animal loco-motion. When you walk or run, you put the right foot ahead of the left after a particular time interval. When you put your right foot ahead, you also swing your left arm forward. The horse does something similar: when its right hind leg moves forward, the left front leg moves forward as well. It's the same for the cat and the mouse. The time step for the cat is shorter, and it is even shorter for the mouse.

Timing is everything, like the firing of the spark plug in an old automobile engine. Timing and fine-tuning do not apply only to mammals. Small birds (such as robins and cardinals) hop less frequently than even smaller birds (such as sparrows and swallows). They are called passerines, from the Latin noun *passer* for small birds. They cling to branches and hop on land. They do not walk like mammals and bigger birds such as pigeons, chickens, turkeys, and ostriches. In order to walk, birds cannot alternate the movement of their "arms" in the same way as quadrupeds and bipeds do. Instead, they thrust their head in synch with each step so that the body's center of mass is always as far in front as possible, helping the body to move forward more quickly. Notice how the pigeon walks and bobs its head, just like the chicken.

Humans perceive *change* in the observed surroundings, not 'clock time'. Perceiving change is so common that each one of us grows up confident that tomorrow will be different than today. The prehistoric *homo* perceived change with moving eyes, not dead scenes with motionless eyes. This is how he knew that "now" is different from "before" and that the future will be dif-ferent from the present. Later, as language developed, he recorded this most common of all observations in words such as *now, before, next, past, present, future, time*, and the passage of time.

Why do we tend to focus on the unusual (the surprise) and not on the ever present? This book unveils the physics behind this tendency. The reason is that the time measured by a clock is not the same as the time perceived by the human mind. The mind time is a sequence of images, or reflections of nature, that are fed by sensory organ impulses. The rate of changes in mental images decreases with age because, as noted already, several physical features change with age: saccade frequency, body size, and pathway degradation. These variations have been well documented in physiology, and they are grouped together as a single phenomenon in Chapter 2.

Beautiful paintings, bas-relief art, and facades of buildings are typically shaped like Figure 1.1, longer horizontally than vertically. They look beautiful for two reasons: the observer has two eyes aligned horizontally, and the eyes *scan* the observed image in short, sudden motions (saccades). Scanning occurs in both vertical and horizontal directions. Repeated scanning, the movement that causes the perception of change in the observed image, is the physical process that underpins the perception of time.

Chapter 3 shows that an image is scanned the fastest (vertically and horizontally) when its two-dimensional shape has an aspect ratio (length/height) of approximately 3:2. Faster scanning means faster understanding of the immediate surroundings. People are unwittingly attracted to such an image because it makes life safer and easier. To move out of danger and to find food, shelter, and mate, faster grasping is essential. This applies to all animals with vision, whether runners, fliers, or swimmers.

Scanning is a movement that unites many features of human life. We scan the earth with our feet, vehicles and communications. Unwittingly, we cover the earth with modular constructs of inhabited areas (block, city, state) shaped such that each area is covered (along and across) the fastest and most economically, just like the human field of vision.

The connection between easy understanding and easily scanned (beautiful) images serves as the physical basis for cognition. Beauty continues to drive and improve cognition. It is precious and indispensable to life and industry, be it art or fashion.

Beauty means attractiveness, the feeling of being drawn to what you see, hear, smell, and touch. The "contrast" present in an image facilitates perception tremendously (Chapter 4). The sharp contrasts between bright colors, and between dark and light shades, are what triggers the impression of change as the eyes scan the image. The sharp differences, like the ruts in the road, shake both vehicle and driver. They are noticed and remembered.

Chapter 4 illustrates how the presence of contrast accounts for several illusionary impressions in the observer's mind. Illusions tend to occur so as to enhance the contrast between regions that touch, when in fact each region is painted uniformly with its own shade. To accentuate the interface, the shade of each region appears nonuniform in the vicinity of the interface. As Harry Houdini put it, "What the eyes see and the ears hear, the mind believes."

Illusions take place while the mind organizes the newly received image among previous images that are already sorted and stored. This is the mechanism of perception. I think of it as the Tetris game, in which a successful move takes place when a new brick falls into a waiting gap, so that the wall of bricks becomes bigger and stronger.

How the Tetris player rotates the new brick to make it fit in the waiting gap is not necessarily how the brick was actually oriented before it arrived. The difference between the two orientations can create the illusion that the new brick was added before it arrived. That is why perception is personal, belonging to the individual. To be in a situation (*in situ*, at a site, position, place, time) is to be in a changing image that you perceive as your "surroundings." At the same time, your neighbors perceive the changing image as their own surroundings, with you in the image that they perceive.

> The world measured, modeled and ultimately predicted by physics is the world of perceptions, a category of *mentation*. The phantasms and abstractions reside merely in our *descriptions* of the behavior of that world, not in the world itself.
>
> Niels Bohr

In this book I ask why the mind "tries" to make sense of a new input. Why is there a natural tendency to organize the fresh input to make it fit among past receptions? The answer that comes from physics is one, and it is general: empowering the individual with speed and clarity of thought, understanding, decision making, and movement on the earth's surface. The same answer holds for the other "disparate" perceptions detailed in this book, from time and beauty to shape, message, perspective, and dreams.

The shapes that attract us are diverse, but not many. Chapter 5 shows that in addition to the golden ratio rectangle (Figure 1.1), prevalent in nature and the human realm are the conical shapes (e.g., hourglass, sand pile, termite mound, teaspoon heaped with sugar), round cross sections and tree architectures in flow channels, and the "convergent" shapes of boats with sails, airplanes, helicopters, pyres, pyramids, and bird-foot-shaped supports.

The perceived image conveys an idea to the viewer. The concept of "idea" comes from the ancient Greek *idein*, which means seeing with the eyes of your mind. Once grasped, the idea is out of the page. It cannot go back. If the viewer reproduces the idea (the mental viewing) and presents it as his own, then that is a stolen idea. Chapter 6 is an illustrated course on how to distinguish between the remake of an idea and the original. The detective work must be done by human eyes because pictures do not lie.

Nature is not a gallery of two-dimensional objects, as in the many illustrations displayed in this book. The third dimension of an object is the "depth" perpendicular to the viewed plane. This is perceived in perspective, which means the view as you look *through* the image. Chapter 7 graphically teaches the method of linear perspective due to Filippo Bruneleschi, a central figure of Renaissance architecture. The method comes from the idea that the animal mind evolved to acquire a simple rule of perceiving and understanding "depth": objects that are near look big, and similar objects that are far look small. The mind compares the big and the small and understands the depth of the image and how close the danger is.

Perceptions of time and beauty are essential in art and science (Chapter 8). The history of technology is recorded as a select parade of objects (devices, inventions) aligned in a particular direction with the passage of time. The same can be said of the history of civilization, for which the history books show us images of much bigger objects: edifices, roads, bridges, and aqueducts. The history of science is its own parade of images, from antiquity to the present. Yet, hidden under the feet of this marching column is the secret of the direction of the march:

In the beginning, the objects of arithmetic and geometry were one-dimensional: lines, segments, size comparisons, and the line axis of numbers. Later, the one-dimensional objects were joined by two-dimensional objects (plane geometry), and then by three-dimensional objects (solid-body geometry). The direction has been toward liberating the form (drawing, design) to exhibit it in more dimensions.

The direction has been toward greater freedom, from one dimension to two and three. More dimensions in designs imply greater complexity in the population of designs. The same evolutionary direction is discernible in the history of art, from prehistoric dots and line markings on cave walls to paintings on the same walls, all the way to the paintings and photographs of the modern era. Then came three-dimensional art, from bas-relief in ancient Middle East to Greek sculpture and, two hundred years ago, French descriptive geometry. One hundred years ago, a new kind of three-dimensional art arrived — the moving picture (cinema) — in which the new dimension was time.

Slow time becomes fast when every train of images (or any activity) repeats itself the same way many times. The opposite is true as well. In April 2020, when the coronavirus lockdown began, many people felt that time had slowed down. They talked and wrote about it. The time slowed down because we were forced to experience new things. The new mode of living brought back elements from before the industrial age. We stayed home, walked around, and got to know our neighbors.

Time slowed down in synch with our own movement. This, however, did not last long. Two weeks later, the new life stopped being new. It had become its own routine. Without warning, time had regained its original fast pace. In the final chapter, I invite the reader to use lessons from the pandemic in order to control the speed of perceived time. For example, to slow down my time, I do several things. I keep my eyes open. I carry a paper calendar on which I see the days of the month, and cross one at a time. I see what lies ahead and what is behind. I write by hand and draw by hand.

I tend to do the kind of work that is not repetitive, not routine. One who is engaged in creative work, such as research, writing, and design, enjoys this advantage. I value each day as if it were my last one. One hears such advice when discharged from the hospital. During unusual events such as crises, fresh ideas, new art, or an exciting football game, we pay closer attention to the changes that occur in the images in front of our eyes, and consequently the brain records more changes per unit time — that is, more changes than when we are bored. The unusual makes us feel that time has slowed down.

Geographical location has an effect on how we perceive the passage of time. The effect of location is due to human activity (movement, speed, fuel burned, wealth, economy, advancement), which is distributed nonuniformly on the landscape. Imagine moving from North Carolina to New York. Human activity is more intense, faster, and full of change in New York.

The moral of the story told in this book is that although scientists may have contemplated aspects of time and beauty separately, it is fun to understand them together, and to predict them. It pays to walk against the marching crowd. That way, you see many more changes than when you are inside the crowd, moving at the same speed and in the same direction as everybody else.

Time and beauty are so obvious and "disconnected" that to put them together in a book of physics requires daring. This book is not only about why these human perceptions are correct and predictable on the basis of physics, but also about why they share the same cause, the same physics. This, the surprising and very simple answer, is the reward to the thinker who questions and has the courage to speak up. The courage to question and to go against the tide is rare and valuable, especially today in the era of big data and "knowledge industry". For me, the questioning learned growing up and at MIT is like what I learned in basketball: you can't beat the training. You can't forget it, that's who you are, that's what you think, that's what you do.

Understanding the physics of human perceptions (time, beauty) brings us closer to the physics of how our mind flows and functions. It ushers the science of form and consciousness. There is physics behind all human preferences. Physics is the human urge to be fast, wise, safe, economical, and all-knowing and to live longer. We learn to get by, we polish, and we keep what works. This, by the way, is evolution as physics. The river basin does it, and so does the urban traffic. We all do it.

Perception is the most empowering feature of the human animal design: vision, hearing, smell, taste, and touch.

The story of science offered in this book is a human story. The history is kept close to the narrative. The physics in the story was first published in peer-reviewed scientific journals [1, 3]. My unusual path in life is intertwined with the history of science and the main plot, which is the emergence of the discipline of the physics of evolution, form, and consciousness.

References

1. A. Bejan, Why the days seem shorter as we get older, *Eur. Rev.*, **27**, 2019, pp. 187–194, doi: 10.1017/S1062798718000741.
2. https://www.ae-info.org/ae/Acad_Main/News_Archive/Why%20the%20Days%20Seem%20Shorter%20as%20We%20Get%20Older
3. A. Bejan, The Golden Ratio predicted: Vision, cognition and locomotion as a single design in nature, *Int. J. Design Nature Ecodynam.*, **4**(2), 2009, pp. 97–104.

2

Speeding time

One of the most common human perceptions is that time passes faster as an individual becomes older. The days become shorter, and so do the years. We all have stories of this kind, from the long days of childhood and the never-ending class hours in elementary school, to the days, months, and years that now pass in a blur. The most common sayings capture this impression: Times flies; Where did the time go? Last year was yesterday; Growing up took forever; A watched pot never boils.

I have been wondering about the physics behind this feeling that most of us share, and here I provide the explanation that comes from physics [1]. More subtle than the speed of years and decades is the impression that some days appear to pass more slowly than others. We do not have to wait till we are old to experience the phenomenon. Slower days are full of productivity, events, and memories of what happened. If you did not notice this difference between slow days and fast days, then you should pay attention to it, because in this finer difference lies the explanation for the lifelong puzzle sketched in the preceding paragraph.

The hint is that days are productive when the body and mind are relaxed, after periods of regular sleep, when in the morning you look in the mirror and see a younger you, not a tired you. Athletes learn the hard way about the correlation between good rest and the speed at which time passes. Lack of rest makes you miss plays, leaving you unable to anticipate, unable to see the ball before it arrives. While sleepwalking, the game is over before you know it.

Students learn the same physical truth while taking exams during a fixed time interval. The rested mind has more time to go through problems, find mistakes, go back to the beginning, and try again. Lack of sleep, due to cramming during the night before the exam, makes time pass fast during the exam period. Cramming does not pay, but rest does. This is why the good parent makes the child sleep before the math exam, and the good coach rests the team before the big game.

Here is why this is important to you, the reader. Today, many young people experience time distortion because they spend too much time on social media. This trend has serious consequences, ranging from sleep deprivation to mood changes, mental disorders, and suicide. This is why understanding the physics behind how humans perceive the passage of time is essential.

Time represents the perceived changes in stimuli (observed facts) such as visual images. The human mind perceives reality (nature, known as "physics") through images that occur in the mind when visual inputs reach the cortex. The mind senses a "time change" when the perceived image changes. The time arrow in physics is the goal-oriented sequence of changes in flow configuration, which is the direction dictated by the constructal law [2–5]. The present is different from the past because the mental viewing has changed, not because somebody's clock rang.

The "clock time" that unites all the live flow systems, animate and inanimate, is measurable. The day–night period lasts 24 hours on all the watches, wall clocks, and bell towers. Yet, physical time is not mind time. The time that you perceive is not the same as the time perceived by another. Why? Because the

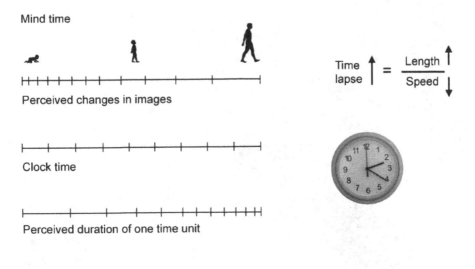

Figure 2.1 The misalignment between perceived time and clock time during lifetime.

young mind receives more images during one day than the same mind in old age. Said another way, if the lifespan is measured in terms of the number of images perceived during life, then the frequency of mental images at a young age is greater than in old age (Figure 2.1). This is why it should be so:

The sensory inputs that travel into the human body to become mental images — the "reflections" of reality in the human mind — are intermittent. They occur at certain time intervals (t_1) and must travel the body length scale (L) with a certain speed (V). In the case of vision, t_1 is the time interval between successive sudden movements of the eye, called saccades (Chapter 3 describes saccades in more detail). The time required by one mental image to travel from a sensory organ to the cortex is of the order $t_2 \approx L/V$. During life, the body length scale (L) increases in proportion with the body mass M raised to the power 1/3, and, like all growth phenomena, the body mass increases over time in an S-curve fashion [6] monotonically,

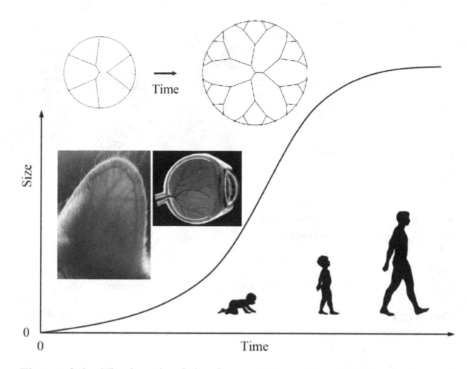

Figure 2.2 The length of the flow path increases as the body size and complexity increase.

slow–fast–slow. An example of the S-shaped history of growth is shown in Figure 2.2. The physics that underpins (and allows us to predict) the S shape will be discussed in relation to Figures 6.1 and 6.2.

The length traveled by the inputs from external sensors to the cortex is actually greater than L, and it increases with age. The reason is that the complexity of the flow path needed by one signal to reach one point on the cortex increases as the brain grows and the complexity of the tree-shaped flow paths increase (cf. Figure 2.2). The broad trend, then, is that L increases with age.

At the same time, the flow speed V decreases because of the aging (degradation) of the flow paths. The key feature is that the physical time (the combined effect of t_1 and t_2) required for one mental image to occur increases monotonically during the life of the individual. The frequency of mental images decreases monotonically and nonuniformly (not at constant rate). This trend is illustrated qualitatively in Figure 2.1. Two summarizing conclusions follow:

- More of the recorded mental images should be from youth.
- The "speed" of the time perceived by the human mind should increase during life. The rate at which the physical time clock "ticks" during one perceived change in the mental image increases with age.

The misalignment of the clock ticks and the changes perceived by the mind brings together numerous observations and measurements accumulated in the physiology literature, especially in the study of vision and cognition. These data support the time misalignment (Figure 2.1). I reviewed this body of literature [1], and here is a synopsis:

During natural viewing conditions, a normal adult subject makes 3–5 saccades in a second separated by periods of 200–300 ms, during which the eyes do not make large or fast movements [7]. These periods are called fixations. If the retinal image as a whole is prevented from moving by successful voluntary attempts not to move the eyes, or by mechanical means, then vision rapidly becomes blurred and the perception of the retinal image eventually fades away completely within 10 seconds.

The highly inhomogeneous structure of the primate retina, with an extremely high density of receptor and ganglion cells in the center (a specialized fovea) and a rapid decline of the cell densities toward the periphery, makes it almost impossible to have a homogeneous and simultaneous percept of the total visual field, without somehow moving the fovea to different positions and acquiring and integrating information from these successive "looks." The existence of a fovea requires both eye movements and periods of fixation, that is, the active suppression of saccadic eye movements.

Although the reaction times of saccades is relatively stable (200–250 ms) [8], the infant fixation times are shorter than in adults. In primates, there is a constant relationship between the duration, peak velocity, and amplitude of saccadic eye movement [9–11], known as the main sequence, in which saccade trajectories evolve toward optimizing the trade-off between accuracy and duration (speed) of the eye movement. This is also in accord with the physics basis for the human preference for displays shaped in "golden ratio" rectangular frames, which is the shape that the two human eyes scan the fastest (cf. Chapter 3). As a result of an interaction between afferent, central, and efferent neural processes, we perceive a complete and stable visual field, which can serve as a frame within which we see motion and within which we move ourselves or parts of our body.

Fatigue can produce overlapping saccades in which the high-frequency saccadic bursts should show large pauses (glissades), in which the high-frequency bursts are much shorter than appropriate for the size of the intended saccades, and low-velocity, long-duration, non–main sequence saccades, in which the mononeuronal bursts are of lower frequency and longer duration than normal [12]. When the saccadic eye movement system fatigues, saccades become slower and the neurological control signal stratagem changes. The term *fatigue* is used here in a broad sense, as a group of phenomena associated with impairment, or loss, of efficiency and skill [13].

The common view that the world is processed as a seamless stream of ongoing perception has been challenged in the current literature. Experimental evidence supports the view that perception might be discrete, further supporting evidence for discrete theories [14]. Visual information processing is similar to a sample & hold mechanism in engineering, as in analog/digital converters. The brain functions in such a way that we consciously perceive only the most plausible solution, rather than a confusing array of possibilities that occur during unconscious processing. The most plausible dovetails with the fastest perceived and with the first impression, and this leads to the human preference for shapes that are perceived more easily than other shapes (cf. Chapter 3).

The period of unconscious feature integration is the duration of sense making. The discrete conscious perception is followed by unconscious processing over time. These two modes of absorbing inputs from the surroundings are analogous to all other flows from one point (e.g., eye) to a volume (e.g., brain) [4]. Observing fast and then letting it sink in slowly is the same dynamic flow design as the "long & fast with short & slow" flows that inhabit all nature, animate and inanimate. Conscious perception and the unconscious processing that follows are the "invasion" and "consolidation" phases of the universal S-curve phenomenon [6], as we will see in the discussion of Figure 6.2.

Experimental evidence casts doubt on the classical model of time perception, which considers a single centralized clock that ticks at a constant rate [15]. The ability to pay attention could modulate the tick rate and hence the duration of the events [16]. Many studies found that the most surprising stimulus (the unusual) within a train is perceived longer, probably because it engages more transient attention or because the event is less predictable. This observation was reported widely during the sudden lockdown at the start of the pandemic, as we will see in Chapter 9. The apparent duration of moving visual objects is greater at higher speeds than at lower speeds [17].

The effect of aging was documented in measurements of horizontal saccades in young, middle-aged, and elderly normal subjects [18]. Peak velocities were significantly reduced in the elderly when target amplitude and direction were predictable. Latencies were prolonged in the elderly under all conditions. Saccadic accuracy was significantly decreased in elderly subjects.

Summing up, the perceived misalignment between mental image time and clock time (Figure 2.1) is in accord with and unifies the growing number of observations that describe physical aspects of this phenomenon in the literature. The physics of this phenomenon is captured by the constructal law of evolution in nature (Figure 2.2). Body growth (the S curve, Chapter 6) and speed of scanning and transmitting the image are the one-two punch responsible for the mismatch between mind time and clock time. As we grow older, the perceived duration of the change from sunrise to sundown feels shorter, even though it is unchanged in clock time.

The young mind matures while getting used to perceive a certain number of notable changes during daylight. The mind does not "count" the number; instead, it "feels" the differences between three impressions: too many, just right, and too few. The human mind is not the only one with this ability. In a recent essay on how animals perceive and navigate on their geography, Kathryn Schulz [19] narrated a situation based on the research of several biologists: "If you trap *Cataglyphis* ants at a food source, build little stilts for some of them, give others partial amputations and set them all loose again, they will each head back to their nest — but the longer legged ones will overshoot it, while the stubby-legged ones will fall short. That's because they navigate by counting their steps."

Ants are not the only living organisms with this ability. Imagine a power outage in your house at night. You are walking in the dark looking for a flashlight. You are not bumping into walls and furniture because you have walked that route many times. You never counted your steps before, and certainly, you are not counting them now. You just know, subconsciously, what is near and far — the feeling inherited from repeated experiences of the same kind. That feeling serves you.

An aging blind man experiences the same feeling as a man with vision. Although the effect of age on saccades is absent, the other effects are present, namely the body growth, the decreased frequency of muscle action in all organs, and the aging of the pathways between the other sensory organs and the brain: hearing, touching, smelling. The sound inputs (think of notes in music) are sampled intermittently, not continuously. Come to think of it, our common impression that the days seem shorter as we get older is rooted in all our sensory organs at the same time, not just vision.

Soon after the publication of "Why the Days Seem Shorter as We Get Older" [1], the physics of the mind time received a lot of attention. In fact, it was the most commented science article published in March 2019 [20]. Many readers sent me comments and contributions to the theory. For example, they pointed out that several theories have been offered to explain the cause of time perception. One recurrent idea is the "proportional theory" proposed by Paul Janet [21] in 1877, according to which for a 10-year-old child, one year is 10% of his life, and for an 80-year-old man it is only 1.25% of his life. So, the year is getting shorter and faster, the argument goes.

Viewed from physics, the proportional theory is similar to caloric theory, but lower in rank because it cannot be tested by measurement. You cannot measure your perception of time and the final length of your lifespan while you are alive. However, it is true that you can measure the amount of "caloric fluid" put into a water bottle by heating it before putting the bottle in your cold bed at night. It is also true that adding one unit of caloric fluid to a warm water bottle is less noticeable to the touch than adding the same unit to a cold bottle in the beginning. The difference is in how that unit is perceived by your touch, not in the number of degrees of temperature by which the mercury rises in the thermometer's capillary tube.

In modern times, what we have in the "proportional theory" is the common-sense adage that "it's never like the first time." Correct impression, good opinion maybe, but where is the physics? What are the facts, the objects? How can you measure them? How can you predict them? Where, after all, is the hard, verifiable evidence? The question "where is the physics?" carries the same weight as Leo Kadanoff's famous question about the alleged fractal "geometry of nature" [22].

By the way, just like the concept of "time," the concept of 'temperature' is derived from human perception rather than instrumentation. *Temperature* comes from Latin: it is the act of mixing (tempering) the hot water bath by adding some cold water. We perceive temperature by observing the height of the mercury column against the scale of the thermometer. Likewise, we perceive time by observing the changes in the height of the column, and this provides a physical illustration of the misalignment between mind time and clock time.

Every hot object exposed to ambient fluid cools down at a rate that decreases in time. The cooling curve is close to an exponential decay curve [23], as illustrated in Figure 2.3. The units of clock time are marked on the abscissa. The units of the perceived time are on the ordinate: the duration of time is perceived as the changes in the number of degrees on the thermometer scale. During the "young" phase of the cooling-down experience, the duration of one unit of clock time is perceived as many changes on the temperature scale. During the "aging" phase of the same curve, the perceived duration of one time unit is short, with few changes on the temperature scale. The difference between the young and the old portions of the curve is that the flow slows down with age. The flow in Figure 2.3 is the heat transfer from the hot object to the ambient: the "speed" of this current decreases with age because the driving force (the

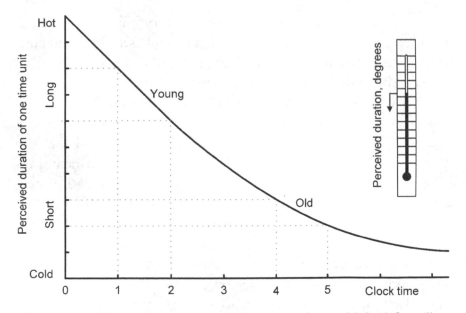

Figure 2.3 The cooling of a hot object exposed to cold fluid flow illustrates the misalignment (Figure 2.1) between perceived time and clock time as the flow becomes older.

temperature difference between the object and the ambient) decreases as the clock ticks.

Viktor Horvath [24] contributed to my article [1] by comparing the functioning of the human brain with that of a computer processor, where one of the most important parameters of performance or speed is the clock. This specifies how many operations a machine can perform per second. If the clock signal decreases, the machine can perform fewer tasks in a given time unit. This can be done differently so that a slower machine can perform a certain number of tasks in more time. This is commonly expressed as "to run faster." The human processor is the brain, and the neurons perform the "count." The neurons communicate with each other through electrochemical processes. We can also say that the number of fires (discharges) per unit time of a neuron is the clock of the human brain. If, as a person ages, his or her neurons can do less work per unit of time, they communicate with each other more slowly, necessarily implying that "everything accelerates in its surroundings" compared to its own juvenile self. Horvath concluded that one does not have to go to the quantum level to explain the oddities of the human sense of time.

Martín Presa, the songwriter of the Argentine band *No Ves Nada*, wrote to me that "for an artist, time seems to stop in a song, a painting, a sculpture... you feel immortal." I feel the same when I come up with a new idea. Creative work unites. Science and art are one burst of pleasure, an avalanche of perceived changes accompanied by the sense that time slows down.

Art and beauty are more commonly known than science, and as a consequence, art and beauty are better instruments with which to demonstrate that science (discovery, knowledge, know-how, and time) is a personal view. The MIT professor Jerry Lettvin taught the psychology course that I took as a junior. He would repeat this gem to the large class in the amphitheater 26–100:

> Remember, each of you is one point in space. You are one point of view, your point.
>
> One point may not seem like much; yet, you are the only one who resides in it, and no one is qualified to displace you.

I was struck by Jerry Lettvin's vision because it placed "the individual" in geometric terms, which are clear, true, and irrefutable. The individual is the stone in the foundation of Western civilization.

I had several great professors during my eight years at MIT, and I remember them as if I saw them lecture yesterday. Surprise, we have just discovered the secret to traveling back in time and how "the time machine" works in each one of us.

To question the physics of time is an important step because time is taken for granted. Scientists tend to focus on the unusual, on change, rather than on what is common and unchanging. Humans have always been subject to gravity, but gravity was not questioned as science until 500 years ago, by Galileo Galilei. Fire too formed an integral part of human experiences for 1.5 million years, but it was not ushered into science until 200 years ago, as "the motive power of fire" by Carnot, Rankine, Clausius, and Kelvin. The phenomenon of mind time is at the same stage of questioning.

Time is the elephant in the room, or, more precisely, in the physics books and classrooms. Today, some authors ask whether entropy and the second law are associated with "the arrow of time." The answer is no [3–5], but the error is understandable. At fault is the authors' willingness to write as if they and their readers have already questioned and now agree on what "time" and "arrow" mean. I know how easy it is to teach that way because I used to teach that way myself.

We are being fooled by our own success, by our standard of living. Time is on everyone's wristwatch and iPhone. It is something that passes by continually; it "flows" independently of our minds, wishes, and daily body condition. That is the major error, and you see it very clearly if you imagine yourself alone, hungry and cold, living in the prehistoric time, when the spoken language was just beginning to empower you. That is when "time" — the idea and the word — was born. The passing of time is the observed change, the new on the mental background of the old. That is the physical meaning of "time arrow".

Advice to the aspiring scientist: Get to know history and art, in addition to science. The following chapters are a few easy steps in that direction.

References

1. A. Bejan, Why the days seem shorter as we get older, *Eur. Rev.*, **27**, March 2019, pp. 187–194, doi: 10.1017/S1062798718000741.

2. A. Bejan, *Shape and Structure: From Engineering to Nature*, Cambridge University Press, Cambridge, UK, 2000.
3. A. Bejan and J. P. Zane, *Design in Nature*, Doubleday, New York, 2012.
4. A. Bejan, *The Physics of Life*, St. Martin's Press, New York, 2016.
5. A. Bejan, *Freedom and Evolution*, Springer Nature, New York, 2020.
6. A. Bejan and S. Lorente, The constructal law origin of the logistics S curve, *J. Appl. Phys.*, **110**, 2011, 024901.
7. B. Fischer and H. Weber, Express saccades and visual attention, *Behav. Brain Sci.*, **16**, 1993, pp. 553–610.
8. K. M. Butler, R. T. Zacks, and J. M. Henderson, Suppression of reflexive saccades in younger and older adults, *Mem. Cognition*, **27**(4), 1999, pp. 584–591.
9. D. Boghen, B. T. Troost, R. B. Daroft, L. F. Dell' Osso, and J. E. Birkett, Velocity characteristics of normal human saccades, *Invest. Ophthalmol.*, **13**(8), 1974, pp. 619–623.
10. C. M. Harris, L. Hainline, L. Abramov, E. Lemerise, and C. Camenzuli, The distribution of fixation durations in infants and naïve adults, *Vision Res.*, **28**(3), 1998, pp. 419–432.
11. C. M. Harris and D. M. Wolpert, The main sequence of saccades optimizes speed-accuracy trade-off, *Biol. Cybern.*, **95**, 2006, pp. 21–29.
12. A. T. Bahill and L. Stark, Overlapping saccades and glissades are produced by fatigue in the saccadic eye movement system, *Exp. Neurol.*, **48**, 1975, pp. 95–106.
13. R. A. McFarland, Understanding fatigue in modern life, *Ergonomics*, **14**, 1971, pp. 1–10.
14. M. H. Herzog, T. Kammer, and F. Scharnowski, Time slices: what is the duration of a percept? *PLOS Biology*, April 12, 2016, DOI: 10.1371/journal.pbio.1002433.
15. G. M. Cichini and M. C. Morrone, Shifts in spatial attention affect the perceived duration of events, *J. Vision*, **9**(1), 2009, pp. 1–13.
16. I. Levin and D. Zakay, eds., Subjective time and attentional resource allocation: An integrated model of time estimation, *Time and Human Cognition: A Life-Span Perspective*, North-Holland, Amsterdam, 1989, pp. 365–397.
17. A. Bruno, I. Ayhan and A. Johnston, Changes in apparent duration following shifts in perpetual timing, *J. Vision*, **15**(6), 2015, pp. 1–18.
18. J. A. Sharpe and D. H. Zackon, Senescent saccades, *Acta Otolaryngol.* (Stockh), **104**, 1987, pp. 422–428.
19. K. Schulz, Why animals don't get lost, *The New Yorker*, 29 March 2021.

20. https://www.ae-info.org/ae/Acad_Main/News_Archive/Why%20the%20Days%20Seem%20Shorter%20as%20We%20Get%20Older
21. https://en.wikipedia.org/wiki/Paul_Janet
22. L. Kadanoff, Fractals: where's the physics? *Physics Today*, September 1986, pp. 11–12.
23. A. Bejan, *Heat Transfer: Evolution, Design, Performance*, Wiley, Hoboken, 2022, section 4.2.
24. https://qubit.hu/2019/01/24/magyar-fizikus-az-idoerzekelesben-szerepet-jatszhat-a-neuronok-mukodesenek-lassulasa

3

Beauty

Have you noticed that you tend to be attracted to things that look nice and are not confusing? The most famous landscapes are painted and hung that way. If you do not visit art galleries, look at the credit cards and bank notes in your pocket. In fact, you are so attached to those things that you carry them with you, and you make sure that nobody steals them from you.

We all do many things unwittingly; yet, there is purpose in good habits. If we question what we think, and especially what we do without thinking, then we begin to see the common thread. This way, we develop a personal understanding of what beauty is and how it works. Beauty is physical, and its usefulness is physical, too.

People prefer images that are roughly half as long horizontally as they are vertically [1]. Photographs, framed paintings on the walls of art museums, cinema screens, computer screens, business cards, credit cards, bank notes, theater tickets, airline tickets, shop signs, flags of nations, and paragraphs are shaped the same way (Figure 3.1). They all have a length-to-height ratio close to 3/2. By pure coincidence, the ratio 3/2 is closer to the irrational number (1.618...) known as the golden ratio, to which I return at the end of this chapter.

Why is this happening, and what does it have to do with physics and evolution? It happens because the image that flows

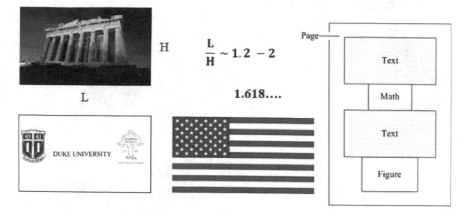

Figure 3.1 People prefer images that have essentially the same shape.

from the retina to brain is *scanned*. The two eyes scan the image horizontally and vertically in short and quick motions called saccades (Figure 3.2). This movement is also responsible for the perception of change, which we associate with the concept of time (Chapter 2).

While scanning in the horizontal direction, the right eye takes over from where the left eye leaves off, and during one time unit (say, one saccade), the length scanned horizontally is longer than the length scanned vertically. In other words, the horizontal scanning speed is greater than the vertical. With middle-school math, one can show that if the shape of the image is free to vary, then the area that is scanned the fastest has the 3/2 shape. With reference to Figure 3.3, the binocular area scanned by the two eyes can be approximated by drawing the rectangular area $HL = A$, where L is aligned with the horizontal (the line of the eyes). The argument proceeds in three steps:

First, the scanning speeds in the L and H directions are V_L and V_H. The speeds are not known at this stage. The time steps required to scan once horizontally and once vertically are $t_L = L/V_L$ and $t_H = H/V_H$, respectively. The time required to scan the rectangular area is $t = t_L + t_H = L/V_L + H/V_H$.

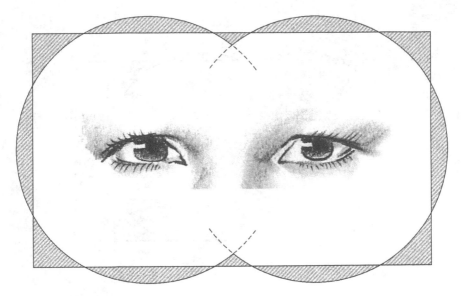

Figure 3.2 The preferred shape is somewhat longer in the horizontal direction, which is aligned with the two eyes.

Second, we ask how the shape of the area influences the total scanning time. The shape is represented by the aspect ratio L/H. The answer reveals itself if in the expression for the total scanning time (t), we eliminate one of the dimensions (say H) by using the area constraint $A = HL$. The expression for the total scanning time becomes $t = L/V_L + A/(V_H L)$. The two terms in this expression change in opposite directions when the shape varies, and as a consequence the time (t) is minimal when the two terms are in balance, $t_L = t_H$, which means $L^2 = AV_L/V_H$, or $L/H = V_L/V_H$.

Third, the conclusion that there is a shape for which the area offers minimum scanning time (namely $L/H = V_L/V_H$) is key, but not the answer. This is because the V_L/V_H ratio is not known yet. We obtain an estimate for the V_L/V_H ratio from the binocular area that was approximated as a rectangle (Figure 3.3). The binocular area is flat-looking, not round, because of the distance between the two eyes, R. During one time unit, the eyes move in parallel and sweep a vertical distance proportional

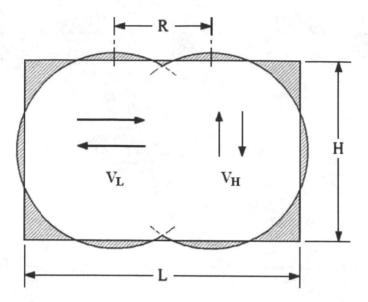

Figure 3.3 The eyes scan the image in both directions, with different speeds.

to $H \approx 2R$. In the horizontal direction the eyes sweep in series, as one follows from where the other leaves off. Therefore, $L \approx 3R$. During the time unit, the effective scanning speed in the L direction is roughly 3/2 times greater than in the H direction, or $V_L/V_H \approx 3/2$.

The conclusion is that the shape that offers easier access to being scanned is roughly 3/2, as shown in Figure 3.4. People prefer this shape unwittingly because it has a physical basis: grasping the image faster is beneficial to animal life. Seeing faster is critical for avoiding danger and finding food, mate, and shelter. This holds true for all the animals with two eyes — the runners, the swimmers, and the fliers. This is also why our "human and machine" species has evolved from humans with lighthouses, sirens, and headlights to humans with radar and GPS.

The same physics argument that led to the 3:2 shape in Figure 3.4 suggests that individuals with only one eye tend to

Figure 3.4 The image is scanned the fastest when its shape is approximately 3:2, which, coincidentally, resembles the golden ratio or the divine proportion.

be attracted to images that look round or square. The one eye is the configuration of the scanner at many checkpoints (airport check-in, theater entrance, store checkout), where the barcode design has evolved toward the square shape. See the upper-left corner in Figure 3.4.

Unlike in primates, cats, owls, and dogs, in many other animals the two eyes do not point in the same direction — forward. In horses, bovines, many birds, and many fish, the eyes are situated parietally, on the two sides of the cranium. For them, the binocular field (Figure 3.3) is more elongated in the horizontal direction, so much that it is pinched in the middle. Such animals perceive the wide surroundings better than animals with two eyes pointing forward but are not as good at seeing directly in front of their mouths. Offer a seed to a hen or a nut to a squirrel with your hand, and you will notice their difficulty. The hen acts as if she is blind. The rooster knows this, and when he finds a seed on the ground, he calls the hen by pecking on the ground repeatedly, to guide the hen to the seed.

For the turbot, the world is the hemisphere above the bottom of the seawater. It has two eyes on its back that point upward. For most of the two-eyed animals, the world looks flat, stretching in three directions (right, front, left), and consequently the eyes are aligned on the horizontal. The eyes of racehorses and horses that pull street carriages are fitted with covers (blinders), which block lateral viewing. This artifact is designed to prevent the horse from being startled by sudden movement from left or right. This is good for the jockey and the carriage driver. Blinders are a relatively recent artifact, from the time when the increasing density of horse-driven carriages made them necessary. There were no blinders on the first domesticated horses and on the horses ridden by the invading hoards.

In summary, the physics of beauty, according to Figures 3.1–3.4, is good for all life. This discovery was tested recently [3] in surveys of several groups of pupils who were presented five rectangular shapes (Figure 3.5). The survey asked, "Which rectangle is the best?" Between 70 and 80 recent

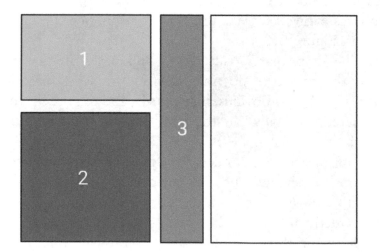

Figure 3.5 Surveys of several groups of pupils were shown five rectangles and asked which rectangle is best [3]. The vast majority of respondents chose shape no. 1, which agrees with the preferences compiled in Figure 3.1 and validates the theoretical shape of Figures 3.2–3.4.

of respondents chose the golden ratio rectangle (no. 1 in the figure). An online mathematics magazine [3] concluded that "we cannot explain this situation mathematically because there is no definition of beauty in mathematics, and the truth cannot be understood by the theory of probability or other mathematical theories. One academic, Professor Adrian Bejan, has found the answer. Therefore, the golden ratio rectangle is more beautiful than the other rectangles and is chosen by the majority of respondents."

Scanning is the movement that defines many aspects of human life on earth. As we saw in the preceding chapter, humans perceive "change" in the scanned image. Time is the idea and word for perceived change.

> "The eye... The whole universe is in it, since it sees, since it reflects."
>
> (L'oeil... Tout l'univers est en lui, puisqu'il voit, puisqu'il reflète.)
>
> Guy de Maupassant

We scan the earth with our feet, vehicles, and communications. Life in the city is a movement on an area. There are many directions, many movers, and immense freedom to change things and ways. There are essentially at least two speeds available to the individual: walking, which is slow and short distance, and riding on a vehicle or animal, which is fast and long distance. This scanning is with two speeds perpendicular to each other, like the scanning with V_L and V_H of any image.

The Atlanta airport serves as an icon for how the urban design takes place, naturally. The shape of the airport area swept by travelers is the same as the shape of the business card. The long dimension is aligned with the fast, and the short dimension is aligned with the slow. On the city block, the two directions are along the street, for riding, and perpendicular to the street, for walking.

A larger urban area is a construct of city blocks. The fast and slow directions are along the avenue and the many streets that feed or draw from the avenue. On even larger areas freight is distributed in the same two ways, long and fast on a few big vehicles on highways, and short and slow on many small vehicles on secondary routes that branch off the highways.

The swept areas acquire their shape naturally from the urge of every individual to have access to the whole area — faster, more freely and securely, and also in a cheaper manner because a single solid infrastructure (street, airport, and highway) serves very many individuals at the same time, and for a long time.

All human movement is of two kinds, fast and slow, and long and short. The area shapes and flow paths that emerge are the floor plan of the building, the evacuation plan for the theater and the stadium, the tall building with many floors and few elevators, the library, the mailroom and package delivery system, the auto assembly line, and the slaughterhouse floor.

Area scanning and shaping are on display throughout the nonhuman sphere — animate and inanimate. The river and the hill slope are the fast and slow of how the rainwater gets out of the plain the fastest. This is why the wetted area is longer in the direction of the river than across it.

The food chain is how distinct animal groups move and live together on the same elongated area. The long and fast direction is traveled by a few large predators, like a few big trucks on the highway. The short and slow direction is for the more numerous and smaller prey, which, in turn, feed on the vegetation and the even smaller and more numerous animals that thrive on the area.

So, that's the physics of why we are attracted to images that we grasp and understand easily and fast. We tend to remember these as "beautiful." We are repulsed by confusing images that slow us down and give us headaches. We put the ugly and scary out of our minds. The soldier returning from the front does not talk about the horror. The famous painter throws away the paintings that do not look beautiful.

Human evolution gave birth to many features that appear attractive and to many concepts that cover such impressions. A beautiful human body is said to be well proportioned. A beautiful face is said to have fine features, which means features that are of comparable size, not dominated by an enormous body part — chin, ears, eyes, nose or jaw. All the preferred objects are proportioned, from the 3/2 shapes of our images to the similar shapes of pyres and the Pyramids, as seen in Chapter 5.

What we can see and draw is the essence of this discussion. We can draw what we imagine and what we can see. On the other hand, we can photograph only what we can see with our eyes or camera.

Proportions are everywhere, beautiful and everlasting. Here is an example from music, courtesy of Victor Niederhoffer [4], who published his ten lessons after reading Fritz Zobeley's book *Portrait of Beethoven* [5]:

The first lesson is greatness. According to Goethe, Beethoven's greatness came from "the feeling for proportions that have a unique beauty and are immortal." To judge his major and minor chords by quantification is as "nonsensical as looking at a painting and inquiring after the chemical formula of the paint used."

Making drawings [2] is an advantage (a mental viewing) in both art and science. The drawing is the permanent record of the fleeting mental viewing. This is why drawings in art and science never go out of fashion. It is also why art and science are one.

When we keep our eyes open, we question. New images strike us, like a slap on the face. They trigger new ideas in our minds and generate unexpected discoveries.

Ideas happen. They are a natural phenomenon of creating new point–volume–point paths in the brain and morphing and enlarging frequented channels in the vasculature of transmission of signals that fills the brain volume. We get ideas when we see things, hear things, smell things, and are surprised by new things.

The new image formed in the mind superimposes itself purposefully on similar images in the mind to make sense of the new as quickly as possible, and to remember and recall the new as quickly as possible. Eye spots painted on the backsides of cattle dramatically reduce the risk of their being killed by large carnivores [6], because the two eyes with the cow tail hanging between them remind the predator of the face of a much bigger and powerful animal, the elephant (Figure 3.6).

The connection between beautiful images and ease of grasping and understanding serves as the basis for the brain design

Figure 3.6   Large predators will think twice before attacking cattle with eyes on their backsides. Photo: Ben Yexley, courtesy of Neil Jordan [6], with permission.

known as cognition. Beauty continues to drive and improve the cognition design. This is why art occurred to cavemen, and why it is now one of the most precious, expensive, and indispensable features of human life, from museums to the fashion industry.

How beauty makes the better brain is documented scientifically in terms of the effect that handwriting has on child development. Handwriting is called "calligraphy" because in Greek the word *kallos* means beauty, and *graphein* means to write. Beauty was in the handwriting from the beginning. Teachers know (and pupils discover) that handwriting is key to understanding better, concentrating better, and remembering longer as one grows up in school. Handwriting improves language comprehension and endows the brain with the ability to access the most important information.

Making beautiful art by hand has the same effect. Parents and teachers are well advised to channel the young toward manual arts of all kinds. Take a hint from the dictum of MIT: *Mens et manus*, mind and hand. Channel the young early, and put a pencil in their hand, not a smartphone. Improvements in communications technology are fine in the domains where they are truly empowering humans, for example, in aviation, security, warfare, and news. In the classroom, the human need is vastly different.

> It is not the walls that make a school, but the sprit that reigns in it.
>
> Dinicu Golescu (1777–1830)

The project of teaching and learning is to form human minds. This requires effort from the receiver, not only from the giver. This is very clear in basketball: if you give a bad pass, do not blame the receiver for dropping the ball. It is the same as in any other sport. The coach can show on the screen all the films and diagrams he wishes, but the better player is not born in that room. The player is born on the court, with the ball in hand and the hunger to become better.

The coach is essential, of course, on the sideline. I learned optimism from my coach: when the tide turned against our team, he would call a timeout and say, "Look ahead, all the game does is wait for a mistake," and he was always right. You can't play defense all day long. You must do both, defend and attack with the proper *balance*, the same way that you inhale and exhale.

The team sport is a live small-scale laboratory for observing social organization. The game is the design (a set of rules), and the sudden opportunities (from mistakes) are the chance and the randomness. Together, rules and chance make the beautiful game.

Mentoring has a role in education. In my professorial career, mentoring was not mentioned much during the previous century. In the past two decades, however, I discovered that during tenure and promotion committee meetings, the discussion became not about the candidate but about his or her colleagues. It meant that the candidate was not as impressive enough because the candidate did not receive mentoring. Amazingly, the committee meeting would become an effort to correct the errors of the professoriate, not the failings of the candidate. In other words, the committee was engaging in group self-criticism (*auto-critique* in French), reminding me every time of my elementary school under communism, in the '50s, when self-criticism was required not from the group but from the little schoolboy who did not sing *The Internationale* anthem loud enough.

In academia today, the duties of the professor include mentoring, in addition to research, teaching, and service. So, what is to be done? The solution is to walk through life with eyes and ears open, because there are good ideas and good role models in your path. In fact, a few of them are extremely well suited for you. The challenge is that the best mentors are the silent ones, the interesting types, the attractive, and the unforgettable.

Observe, learn, get inspired, and this way you hear your *own* calling. That is the secret to perennial happiness in creative work and the only path to silencing your judges. Make the effort.

It takes two to mentor.

All the things that we are taught in school are often described as beautiful: mathematics, physics, zoology, botany, geography, music, languages, and, of course, art. The danger that the smartphone and the internet pose today was obvious many decades ago, when lecturers of all stripes walked on stage with transparent foils, projectors, and laptops instead of chalk in hand. They did more than claim a superior method. They claimed that the future of the teaching profession belongs to their high-tech approach.

Today, most of the classrooms at Duke are for handwriting and drawing on a wall, because, as in all evolution in nature, what works is kept. Smaller effort from the body of the giver equipped with high-tech does not imply a superior brain in the body of the receiver.

Evolution empowered animals to grasp the landscape and its message more easily and faster. The emergence of the two eyes aligned horizontally is the most obvious indicator of the cause. Another sign is the emergence of the two ears, also aligned horizontally.

At the same time, evolution has endowed animals with features that make them easier to be perceived by their kin. The big size is an advantage: when under threat, the bigger animal can strike with a greater force. Features that make the animal appear bigger are the ruffled feathers on the rooster, the bear raised on its hind legs, and the big eyes on the many birds, insects, and cattle. Evolutionary design empowers an animal to attract a mate and to fend off an attacker.

Evolutionary design also endowed animals with features that make them difficult to be grasped by predators. This is the physical basis of camouflage, which takes many forms: fur coat that blends in the grass, padded feline feet for noiseless walking, and owl feathers for noiseless flying. On the other side of the battlefront, the evolved design of the predator is to see and hear the prey, and not to be seen and heard.

The physics of beauty shines a light on the relationship between religion and science. During the Renaissance, the shape of the attractive images seen all around was named "divine proportion." It was attributed to God because an explanation from physics was not available. Beauty, in its many appearances and places, was thought to be God-given. Animals were named "creatures" because they were thought to have been created by God.

That was before science spread. With science, the record of observed facts remained essentially the same, but knowledge became simpler and easier to remember, to teach and to rely upon in order to predict, to anticipate, and to put to human use.

Science came as an add-on to religion. Science and religion are two artifacts, and both were adopted by humans to make life (movement) easier, safer, and longer-lasting. Like the ancient religions, science is evolving. Science strives toward its own "monotheism": the theory of everything, the law of all laws, the fewest laws for each discipline, and the "Einstein icon" of intelligence that will never be replaced by another icon.

Religion is older, yet it continues to be useful in science. Scientific language borrows from the language that was before science — for example, in the organization of science (hierarchy, ceremony) and in the nomenclature of academia (dean, chancellor) and dress code (cap and gown). The scientist who predicts is like a 'prophet'. The correct prediction is met with 'belief' and then used and followed. The embraced science becomes state science, paid for and enforced by the state, just like embraced religion.

The ethics of scientific publishing (do not lie; do not steal) are rooted in Judeo-Christian teachings. Biologists speak of animal and animate, which originally meant the soul (*anima* in Latin). The most respected voice of science in nature documentaries these days, David Attenborough, speaks of "creatures." Nobody is accusing him and his guests of being creationists.

Nature speaks to us in the language that she taught us already. The human mind has the natural urge to understand faster, which means to rationalize, simplify, and explain what it needs to remember, retrieve, and use more easily.

The mind stores the imagined and the unseen in the old language of the observed and the touched. This is why analogy springs in the mind, and why analogy is automatic and useful. This is how humans became empowered with thought, superstitions, religion, and science.

Beauty is usually associated with looking good. This is incomplete, and often misleading. Beauty means attractiveness, the feeling that you are attracted to what you see, hear, smell, touch, and use. Beauty is also associated with an intelligent statement, idea, or person. Attractiveness, comprehensive as it may sound, is not goodness. The record shows that bait, poisonous mushrooms, and Hitler's marching music are attractive.

Beauty is attractive because it is *useful*, physically. The features and habits reviewed here happened naturally in the long run of human evolution because they made our ancestors better, more efficient, safer, longer-lasting movers on earth. The individual wants to be left alone in peace.

References

1. A. Bejan, The Golden Ratio predicted: Vision, cognition and locomotion as a single design in nature, *Int. J. Design Nature Ecodynam.*, **4**(2), 2009, pp. 97–104.

2. A. Bejan, *Freedom and Evolution: Hierarchy in Nature, Society and Science*, Springer Nature, Switzerland, 2020, chapters 6 and 7.

3. MATHVN.COM, Why is this rectangle the best? (Tại sao hình chữ nhật này lại đẹp nhất?), Vietnam, 13 October 2020. https://www.mathvn. com/2020/10/tai-sao-hinh-chu-nhat-nay-ep-nhat.html.

4. V. Niederhoffer, 10 Lessons from the life of Beethoven, *Daily Speculations*, 18 December 2013.

5. F. Zobeley, *Portrait of Beethoven: An Illustrated Biography*, Herder and Herder, New York, 1972.

6. C. Radford, J. W. McNutt, T. Rogers, B. Maslen, and N. Jordan, Artificial eyespots on cattle reduce predation by large carnivores, *Commun. Biol.*, **3**, 2020, p. 430.

4

Contrast

The images that nature shows us are not like those that people make. They are not even like those that the most gifted artists make. People make images by drawing lines on a uniform background, usually white. This happens in both art and science. Later, if there is need and talent, the image maker fills the areas between the lines with colors and shades. Even later, the best of the artists blur the lines or even paint over them, leaving only the nonuniform color that fills the page. Today, thanks to high-tech photography, experts marvel when they discover the line drawings hidden under the most famous paintings for many centuries.

Why does it happen this way, in the evolution of image making from the earliest and simplest cave drawings to the masterpieces that continue to inspire?

The reason is that nature is not made of line drawings. The images of nature fill the field of vision completely, not partially. Nature is not fractal. Every image is full, a continuum with distinct features distributed nonuniformly. The colors change from spot to spot, and so do the shades. They are what the great masters were after, from Michelangelo to Matisse. Nature's images strike us with the differences between spots and their colors. Nature strikes us with contrast.

In human perception, contrast is the same as the change in time, a topic discussed in Chapter 2. When the eyes scan the field of vision in saccades, the change (in color or shade) from one spot to its neighbor is perceived as a change in time. This change is worth observing, transmitting, and recording in the brain. It is the same phenomenon as recording the change that occurs in the image when it is morphing or moving.

Shape, structure, and message in the image are names for the ways in which the nonuniform spots are arranged in front of us. The arrangement is with meaning. The shape of one spot is discernible from the local background constituted by its neighbors. The eyes discern the shape, and if there are several discernible features of this sort, the eyes also discern the construction and message in nature's image.

To discern or not to discern, that is perception. The word *discern* says it all. It combines two mental viewings, from the Latin *dis* (apart) and *cernere* (to separate). This viewing happens while sifting the flour, panning for gold, and distilling gasoline from crude oil. Discerning needs a design — brain, sifter, pan, and distillation column — where the last three designs come from the brain as well. More precise than "to discern" is "to distinguish," which comes from another mental image in Latin, *dis* (apart) and *distinguere* (to prick, to differentiate), as if you are using a needle or a stinger to point to what is important. To pick the olive, you need a toothpick.

As we saw in Chapters 2 and 3, the evolutionary design of perception (not only in humans) followed its natural course toward a faster, clearer, and less ambiguous grasp of the observed. This is why the receiver of the perceived image — the brain — has evolved in the direction of simplifying and condensing the message in the image. How was this accomplished? You will know the answer if you ask yourself how you would draw nature's image if you had little time to get out of the way of danger. You would simply draw lines between the spots that mesh but look different.

The mind has evolved such that it enhances the contrast between neighboring patches that are related but not the same. This is a great advantage that we all share unwittingly. Just look at the upper half of Figure 4.1. Four rectangular areas are positioned flush together, and in each area the shade becomes lighter from left to right. This impression is an illusion [1] because the shade of each area is uniform, as shown in the lower half of the same figure.

Why the illusion? Because it is a lot easier and faster to understand that there are four different shades by seeing the contrast between them. Illusion, like gossip, can be useful.

Figure 4.1   In the top row, each of the four areas gets lighter when scanned from left to right. The contrast between two touching areas is sharp. Yet, on each area the color is uniform, as shown in the bottom row. The human mind enhances the small differences between images (or colors) that are similar but not identical.

In this case, nature shows us "boundaries" without the presence of thin lines between adjacent areas that are only slightly different. It is as if the lines were erased or covered with the shades that were painted later.

By the way, illusion does not happen if the top panels in Figure 4.1 are of different colors. The human ability to accentuate the contrast as the eyes scan across the boundary between panels is not needed because the contrast is present already in the different colors.

A mathematician would be correct to argue that vertical lines are indeed present between the four areas in Figure 4.1 (top), because in geometry a line has zero thickness. We see the beauty of geometry in such an argument, because the impression of beauty (cf. Chapter 3) comes from being able to understand an image faster. The zero-thickness line of geometry is an abstract concept, valid in the limit of an infinitely thin line that one thinks of sketching. For ease of teaching and learning, in geometry we draw thin lines with pencil, ink, and chalk. These lines have finite thickness, but we get on with the argument because both teacher and student agree on the zero-thickness definition of "line." On the other hand, lines with zero thickness would be impossible to see in Figure 4.1 and anywhere else. To argue that they are present is akin to admiring the emperor's new clothes.

Thin and light lines are very useful in making a drawing, especially in making the first moves on the page. We see them in many figures in this book, for example in Chapter 6, where the invasion path that initiates the spread of a flow on an area (Figure 6.2, left) is a thin line on paper. In nature it is not a thin line. The river channel has its riverbed, which has width, color, and shape that meanders or not, all of which call for the eye, talent, time, and patience of the artist.

Not all line drawings have lines that are really thin, or lines that are equally thin. A natural image approximated (modeled) as a thin-line drawing has at least two length scales, the *external* length scale (the page), on which the image is presented, and the *internal* length scale, which is the average thickness of the lines. The ratio of the external and the internal length scales is the svelteness (Sv) of the piece of art [2], and Sv is a characteristic number greater than 1. Among other features, the number Sv distinguishes between the original drawing and the remakes, and it is useful in assessing the originality of new art. In Chapter 6, we return to the effort required to distinguish the remake from the original.

There is a lot more to what Figure 4.1 reveals about the human mind. The magnification of contrast between closely similar impressions that are not identical is present in many other acts of perception. We remember the contrast, not the deluge of information. We remember the two-word titles of thick books and long movies: *War and Peace, Beauty and the Beast, The Many and the Few, The Red and the Black, The Prince and the Pauper, Crime and Punishment,* and *A Fistful of Dollars.* In the natural sciences, we frequently read about "the physics of life," in spite of the fact that in academia today physics and life are separate fiefdoms. Life is the stuff of biology, whereas matter (the inanimate) is the stuff of physics. We read more and more frequently of "design in nature" even though current teachings associate "design" with direction (objective, purpose, and human involvement) and "nature" with the nonhuman realm — random changes, mutations, and chance. In poetry, authors of haiku achieve an unexpected effect by juxtaposing two different subjects, one human and the other from nature.

Difference is interesting, uniformity is not. This is particularly true in scientific work. We tend to notice and retain the differences, not the unifying features. This human tendency is the mother of all stereotypes, from how faces look, to accents and behavior.

Colors and shades are everywhere in nature and art, while scientific drawings have thin (sharp) lines. Blurred lies are a superior feature for grasping and communicating the outlook of nature. Sharp lines are primitive and misleading, not part of nature. In the science of thermodynamics, we see the shortcomings of drawing a line (called boundary, incorrectly) to separate the spot of interest (called system) from the rest (called environment). Because the drawn line has a finite thickness (Figure 4.2), it is mistaken for the wall that surrounds the system.

In reality, the wall is itself a system, as shown in Figure 4.3. The confusion continues to propagate (in spite of the correction [3]) because, thick or thin, the wall in the drawing does not violate the first law of thermodynamics. As shown in the upper part of Figure 4.2, when there is a temperature difference (T_L, T_H) across the wall, the flow of heat (Q) is conserved from right to left. The fact that the wall is not the boundary intended

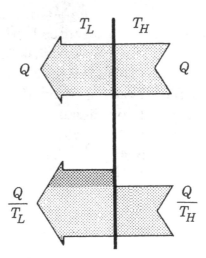

Figure 4.2 The wall drawn between two regions (systems) at different temperatures is not a boundary. The flow of heat (Q) is conserved though the wall, but the flow of entropy (Q/T) is not. The wall is itself a system, which in this example is the source of the irreversibility (one-way flow) of the assembly [2].

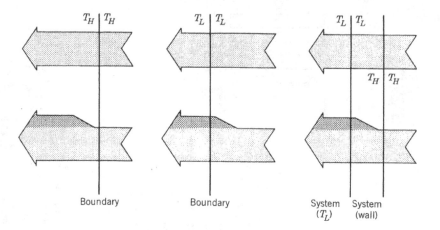

Figure 4.3 The difference between drawing a boundary and a wall. Across a proper boundary (a line with zero thickness), the temperature is continuous, and the flows of heat and entropy are conserved [2].

by the scientist is shown in the lower part of Figure 4.2. The flow of entropy (from Q/T_H to Q/T_L) is not conserved across the wall (Figure 4.3), because the wall is the overlooked system that generates entropy, or irreversibility, which is measured as the difference $Q/T_L - Q/T_H$. The wall has a finite volume in which heat flows *irreversibly*, along a temperature gradient.

Illusions that come from the human tendency to accentuate contrast are common. Take a casual look at Figure 4.4, and your first impression will be that the upper drawing is taller than the lower. In reality, both are square — the same square with stripes. The lower figure was rotated 90 degrees relative to the upper. Why this impression? The reason is the same as explained in Chapter 2. In the upper image, more frequent changes are observed when scanning in the vertical direction. In the lower image, the inputs are more "interesting" from horizontal scanning because the eyes perceive more changes in that direction. The difference between the number of perceived changes vertically vs. horizontally creates the impression that

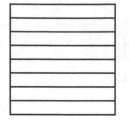

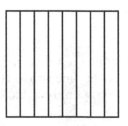

Figure 4.4   Two striped squares are perceived differently, depending on the orientation of the stripes relative to the horizontal alignment of the eyes. The upper image seems taller than the lower. The reason is that scanning in the vertical direction makes the eye perceive changes more frequently, giving the impression that it takes longer to scan.

time slows down in the upper square while looking in the vertical direction, which means that the vertical dimension in the upper square seems longer than in the lower square.

It is similar to driving your car over a surface with parallel ridges. Such surfaces form naturally after intense driving on snow, gravel, and mud. You remember when you drove perpendicularly to the ridges rather than along them. The mechanical integrity of your vehicle remembers it too.

Contrast is the source of illusion in Figure 4.5. Each drawing uses the same two colors, blue and green. The spot in the center is a fine "chess board" of tiny blue and green squares present in equal numbers. Both drawings have the same center spot. When the background is blue, the spot looks green. The right side of

Figure 4.5 Two colors, blue and green, on two drawings. The spot in the center is a 50–50 mixture of the same two colors. On a blue background the spot looks green, and on a green background the same spot looks blue.

the figure gives the opposite impression: on a green background, the spot looks blue.

I recall becoming curious about these contradictory impressions when I was a little boy. My mother had green eyes with blue speckles. I was intrigued by the fact that her eyes seemed green when she wore green and blue when she wore blue. It was as if her eyes were reflected in her blouse, and vice versa. I kept having similar impressions ever since, in different colors. Most surprising is the effect of a red or pink blouse, which brings out the pink that is present in the eyes, especially after lack of sleep. Guy Marks made this impression famous in his 1968 song "Your red scarf matches your eyes."

Figure 4.6 recreates the impression from my childhood. When you look furtively at the whole figure, the center spot looks blue on the blue background. This seems to contradict the impression derived from Figure 4.5 (left), where the spot stood out against the background. There is no contradiction, because the white "cornea" separates the center from the background. From the point of view that covers the whole, there are two colors, blue and white, and the contrast between them. Contrast means that everything that departs from the

Figure 4.6   The color of a person's eyes mimics the color of the blouse. In this montage, the center spot looks like the blue background even though its true color is the 50–50 mixture of blue and green seen in the center dot of Figure 4.5.

leading minority (white) is relegated to the majority (blue). This is the contrast, the dichotomy, the island with white sand in the sea of blue. The inverse of this demonstration is shown in Figure 4.7, except that one should look at this figure afresh, after forgetting about Figure 4.6.

Contrast is not only perceived in the difference between orientations (Figure 4.4) and colors (Figures 4.5–4.7) but is also brought into evidence by the shape — the arrangement — of the two colors. Green and red are the two colors in the two examples in Figure 4.8. On the left, the two colors are distributed relatively uniformly, and consequently the viewer is struck by the beauty of the whole image, which by the way has the 3:2 shape predicted from the urge for access in Chapter 3. On the right, the red and the green are arranged in such a way

Figure 4.7 The reverse of the colors of Figure 4.6 reinforces the impression that the center dot mimics the color of the background. The center is the 50–50 mixture of blue and green used in the center of Figure 4.5 (left and right), and the background is green. Consequently, the center spot looks green, unlike in Figure 4.6, where the same spot looked blue.

that one red flower stands in contrast to the green. The beauty in the contrast attracted the bumble bee and the photographer.

A similar comparison is offered in Figure 4.9. Animal tracks are usually overlooked, unless one makes a living as a tracker on a safari vehicle or as a hunter. Such occupations are rare these days, but as we will see in Chapter 9, the busy person interested in slowing down the flow of his mind time is well advised to look all around and observe the movement, the changes, and life. Animal tracks are part of the discovery — the other part are the animals themselves.

On the left in Figure 4.9 we see the tracks very clearly, because of the stark contrast. The loops in the trace left by a stiletto fly larva clash with the parallel arrangement of the sand

Figure 4.8   Contrast is perceived not only in the difference between colors but also in the difference between the shapes in the arrangements of the colors. The green and red leaves on the left are uniformly distributed through each other: their diversity is easily overlooked, and the whole derives its identity from the nearly perfect mixing. On the right, the color red stands out in the center of the green surroundings, and its identity draws the attention of insect and artist.

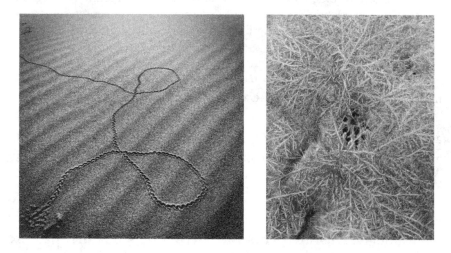

Figure 4.9   Contrast versus shape: Animal tracks with contrast (stiletto fly larva at Oregon sand dunes) versus animal tracks with shape (mink paw in frozen sand on riverbank at Oxbow Park, Oregon). Photographs by Garth Olson and Willem Larsen, with permission.

dunes. On the right, the first impression is uniformity, not contrast. All the details have nearly the same shape and length scale. Only the tracker's eye seeks what is hidden in the frozen sand: the paw of a mink, in the center of the photograph. The toes outline its shape. There is a second paw print, to the upper right, but it is not as clear. Disorganized as it may seem, the frozen sand strikes us with its own contrast: the white needles, branches, and constructal trees of ice on the dark background.

The light shirt color on the dark background of the playing field is a proven advantage in sports. Peder Zane and I documented this advantage during the 2014 soccer World Cup in Brazil [4], when we wrote:

> The white flag is the sign of defeat in battle. The white towel thrown in the boxing ring conveys the same message. Yet lighter colors, especially white, are actually the mark of champions in almost every sport. For example, 83 percent of the winners after the group stage at the World Cup in Brazil wore lighter colors than their opponents. This is neither an anomaly nor a coincidence. Instead it reveals a deciding factor in all team sports that, funnily enough, is recognized but not seen. This factor provides surprising insight into the phenomenon of "home-field advantage" and how subtle physical advantages rooted in physics often mean the difference in closely contested matches.

The conclusion is that nature's image speaks to us through contrast, and when the contrast is not immediately evident, full of small components all mixed together, nature alerts us through the *shape* of the arrangement. This conclusion is a natural transition to the next chapter, which is about the surprisingly few shapes that we know about, because they are all around us.

References

1. S. Vogel, *Life's Devices*, Princeton University Press, Princeton, NJ, 1988.
2. A. Bejan, *Freedom and Evolution: Hierarchy in Nature, Society and Science*, Springer Nature, New York, 2020, Chapter 6.
3. A. Bejan, *Advanced Engineering Thermodynamics*, 4th ed., Wiley, Hoboken, NJ, 2016.
4. A. Bejan and J. P. Zane, The World Cup's light-color advantage, *The Daily Caller*, 7 July 2014.

5

Shape

Shape is form, and the form is chiefly responsible for conveying the idea that is in the image. The word *idea* comes from the ancient Greek word *idein*, "seeing" with the eyes of your mind [1]. Form is not everything. Contrast also helps because it delineates the form and facilitates the transmission (Chapter 4). How well the shape of the image fits inside the natural shape of the observer's field of vision helps as well. The match between the two shapes facilitates the transmission, and for this reason the image feels attractive and beautiful (Chapter 3). Better perception of shape and more changes that are perceived per unit time stimulate the feeling that time slows down and life is longer and more productive (Chapter 2).

Shape, contrast, and change are the natural features of an image. They are manifestations of the same physics principle of evolution of design in space and time. These features occur naturally and are perceived as good, like the new shoe that fits perfectly on the foot. In this chapter, we focus on shape itself and, in spite of the common impression that shapes are extraordinarily diverse, discover that the shapes that matter are not that many. They are few, like the few drawings in a good book of elementary geometry.

This book began with the perception of changes in configurations, which is the origin of the concept of time in the human

Figure 5.1 The conical shape of the sand pile unites the tiny hourglass with the big construction site and all the forms of relief (hillslopes and long valleys) that define the land surface. Viewed from above, the channels dug by the rain look like a round river delta that flows radially and dendritically outward.

mind. That is why we begin this chapter with the hourglass (Figure 5.1, left), which is one of the two oldest instruments (1600 BC) conceived for the purpose of measuring a time interval independently of the person who uses the instrument. The other old instrument is the sundial (1500 BC).

In the hourglass, the measurement of time is an observation of change. The time measure corresponds to the change in the location of the small amount of sand, from the upper chamber to the lower chamber. Three designs of hourglass are shown, each with the same amount of sand but with neck diameters that increase from left to right. The time interval indicated by the complete emptying of the upper chamber decreases from left to right by design, from 5 to 4 and 3 minutes.

The shape of interest here is what happens naturally inside the hourglass. The conical surface of the sand pile in each of the lower chambers is natural. The shape is the same from left to right, and it is time independent. The conical surface advances upward as the sand keeps falling. It is unmistakably the shape of a "sand pile," small or large. The right side of Figure 5.1 shows the same shape during the growth of a sand pile outside a quarry. If you are not familiar with the hourglass and the quarry, try playing with a teaspoon in a bowl of sugar. You will always see a conical pile of sugar in your teaspoon.

From the hourglass to the quarry, the diameter of the sand pile changed from 2 cm to several meters, but the conical shape remained the same. Why is the conical surface inclined relative to the horizontal surface? The reason is that the flow of sand down that surface must be driven by a force, which in this case is the surface-aligned component of the weight of each grain of sand. This force is necessary so that the grain can overcome the opposition (the obstacle) represented by the older grains that make up the rough surface of the cone.

The shape of the sand pile is more than an example that unites the small with the large. It is an icon of the forms of relief that constantly redefine the surface of the earth. Avalanches of rocks, snow, and fluidized wet soil (landslides) are large-scale versions of the same flow of solid as in the sand pile. All these flows are driven by gravity against the friction due to the superficial deformation caused by the flow on the inclined surfaces.

The rain that falls on sloped surfaces contributes additional features to the relief, and the surfaces are even better known for these secondary features. The rain flows along the surface in two distinct ways, by seepage through the inclined porous layer and as rivulets in open channels dug into the surface. Valleys and hillslopes emerge naturally on the incline. Valleys branch out into more valleys toward the apron of the inclined surface (Figure 5.1, right).

Viewed from above, the sand pile and the rivulets resemble a circular river delta with the source in the center of the disk and the sinks distributed all over the wet cone and its bottom perimeter. Near the top, the valleys are deeper and narrower than near the bottom of the cone. This variation is the same as on any mountain, where the narrow and deep valleys with fast streams become wide and shallow with lazy rivers as they approach the plain.

The conical shape and its erosion are evident in termite mounds, in southern Africa (Figure 5.2, left). Termites continuously bring soil upward, to the surface, and the rain and wind do the rest. Solar irradiation contributes to the sculpture, as termites bring more soil to the irradiated side of the mound [2, 3]. The anthill is a more familiar small-scale version of the shape and the construction phenomenon (Figure 5.2, right).

If the conical surface is concave and facing upward, like a funnel, as in a glacier caldera, then the avalanches and their rivulets exhibit pairing (not branching) and converge from the conical surface to the center. Viewed from above, the funnel-shaped surface looks like a circular river basin for which the water source is the wet disk, and the sink is the lowest point, in the center. In Figure 5.3, this natural configuration is made visible

Figure 5.2 Termite mound in Lower Zambezi National Park, Zambia, and an anthill built against the sidewalk near my home.

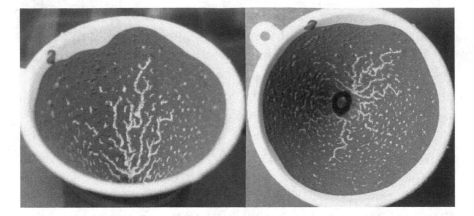

Figure 5.3 When the conical shape is facing upward and its surface is erodible, the channels converge in tree-shaped fashion toward the center. In these photographs, they are visualized with wet coffee grounds painted on the surface of a funnel. Viewed from above, the channels resemble a circular river basin that discharges itself to the center.

by coffee grounds left as residue on the upward-facing surface of a funnel [4].

River basins and deltas are icons of another shape that is everywhere: the shape of the tree, which is also known as arborescent and dendritic (from the word *dendron* for "tree" in ancient Greek). This configuration unites the animate flows with the inanimate flows — vascular tissues, snowflakes, lungs, lightning, city traffic, vegetation, underground rivers, termite mounds, and global air traffic.

The tree flow configurations emerge naturally from their common tendency to evolve freely toward greater access between a point and an area or a volume, in accord with the constructal law. Many classes of tree architectures have been deduced from the constructal law, and many tree designs have been implemented in technologies for heat and fluid flow, transportation, and pedestrian movement.

The oneness of natural tree-shaped architectures in the inanimate realm (river basins, snowflakes) and the animate realm (human lungs, city traffic) is evident and intriguing.

Recent advances are drawing attention to the phenomena of evolution that are thought to belong most generally to physics [4–17]. All such phenomena are predictable. Examples include the treelike shapes of lungs [18] and corals [19], the life span and life travel of animals, vehicles, rivers, and the winds [20], and the arrow of time of the evolutionary organization [21].

The round cross section of a duct is another shape that unites the inanimate and animate flow systems. Round tubes are a common presence in most tree architectures, from blood vessels and lung airways to the invisible "soil piping" that brings the rainwater from the wet hillslope to the rivulet in the valley. You find water pipes in wet soil when you dig for earthworms to go fishing in the nearby river. Underground rivers are everywhere, and their cross sections are always roundish, not flat. Most are unknown; if known, they are out of sight unless they dried up a long time ago and were left for us to admire the vaulted chambers of famous caves. Underground rivers emerge regularly under paved streets in the city. The "sink-hole" phenomenon is how they reveal their presence. On the left side of Figure 5.4, the underground flow is in the forward direction, toward the low-level park in the background. The caved portion of the asphalt reveals a channel that is as wide as it is deep. The cavity is the size of a small automobile. The direction of the channel is forward and downhill, as indicated by the major cracks under the photographer's feet.

The underground river is just one manifestation of the water seepage and erosion that go on under the pavement. The phenomenon occurs on a scale as large as the entire pavement, and it is incessant. Rain falls regularly, and a significant fraction of the water flows straight down, into the cracks that are visible on the pavement on both sides in Figure 5.4. Slowly, this downward flow has the effect of compacting the filling material (clay, gravel) that supports the pavement, and as a consequence, the

Figure 5.4 The sink-hole phenomenon reveals the presence of the underground river. In the photo on the left, the river runs downhill toward the low-level park in the foreground. The 1.5 m depth of the channel matches the width of the hole in the pavement. On the right, the pavement cracks in a tree-shaped fashion when placed in tension by the slow-sinking regions far from the center of the photo.

pavement sinks slowly over time. It does not sink at the same rate because in certain spots the substrate is more resistant to being compacted by the vertical seepage of water. A bulge appears over these spots, as on the right side in Figure 5.4. The sheet of asphalt is in tension, like a soap bubble, and it relaxes its stress by cracking in a tree-shaped pattern.

The big cracks are few, and the small cracks are many. They constitute their own hierarchy. Anybody can see their organization by walking and looking at the ground. To get an idea of the scale, note that the area photographed on the right side in Figure 5.4 is roughly 0.5 m × 0.5 m, and the height of the bulge (the center of the figure) is roughly 1 cm relative to the peripheral regions of the image. The cracks hint that the flow under the pavement carves its own river with a roundish cross section.

Incidentally, the hierarchical cracking illustrated above is the explanation for the "earthquake" phenomenon and for the difficulty to predict its occurrence, the time, and the place. In the asphalt example, it takes time for the bulge to emerge and for the first cracks to appear. This initial period of stress buildup depends on the sinking that goes on elsewhere, far from the bulge. In other words, the cause of the bulging and cracking is obscure, invisible to the observer who focuses on the first crack, which means the first small tremor, the first noise.

The tremors are hierarchical, like the cracks, and are recorded as a broad spectrum of frequencies. While small snaps trigger high-frequency vibrations, big snaps trigger low-frequency vibrations. One can verify this by bending, snapping, and comparing dry pieces of wood, toothpicks, pencils, and baseball bats. The seismologists do not have access to places and images such as Figure 5.4 (right). Even if they knew where the snapping is triggering tremors, the dendritic cracking would be three-dimensional, in a volume, time-dependent, and strongly influenced by nonuniformities in the bedrock and its mechanical properties. In other words, the earthquake phenomenon is a lot more complicated than the cracking asphalt. This is why the earthquake is obscure. At the same time, its origin is as simple and clear as in the bulging asphalt that cracks hierarchically in front of your feet.

Above ground, the round cross section is hidden in every river, large or small. The cross section is shaped as a lemon slice, with a horizontal segment on top. It is shaped this way because of gravity, which makes the upper surface flat and "free," with negligible shear forces between the flowing water and the air above. Generally, the lemon slice has two dimensions, the width

(w) and the depth (d). The cross section of the river is special because its width is predictably proportional to its depth (cf. Ref. 4, section 13.3.5), implying that it has only one dimension, like any other round cross section with friction distributed continuously over its perimeter.

The natural birth of the round cross section can be witnessed in real time by looking at the upward development of a plume of smoke. Two examples of turbulent plumes are compared in Figure 5.5. On the left, a flat curtain of smoke rises from a row of smokestacks. On the right, the plume comes from a concentrated fire. Above an altitude roughly ten times greater than the width of the source of smoke, the two plumes have the same time-averaged cross section, which is round. This behavior is significant because the original shape on the left is flat, and on the right it is round. Said another way, the flat cross section morphed to become round, but the round cross section of the plume on the right did not morph to become flat [22].

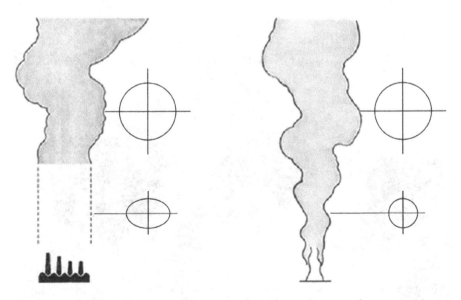

Figure 5.5 Flat plumes and jets morph and acquire round cross sections. Round plumes and jets maintain their round cross sections.

The natural tendency toward the round cross section is due to what flows in the plume (or jet) configuration. The flow is the lateral movement (called momentum in physics, from the Italian noun *movimento* for movement, or *mouvement* in French) from the rising column of the fluid to the surrounding fluid medium, which is stationary or slow-moving. The transfer of movement is perpendicular to the moving column, and the transfer (or mixing) between column and ambient takes place faster when the column has a round cross section.

Many flow systems that are not configured as trees have a round shape. The spreading flow is extremely common. Figure 5.6 shows how ink droplets spread on a horizontal piece of paper. They spread differently, depending on the altitude from which the droplet is released. The first spot is a round "splat" because its droplet fell from close to the paper. The third spot is a "splash" because the droplet was released from farther up and was falling faster when it hit the paper. The spot in the middle fell from an intermediate altitude, and its shape is intermediate as well.

Two observations are worth retaining. Broadly speaking, all three spots are "round": the spread is a flow that proceeds

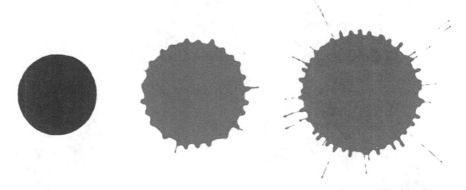

Figure 5.6　Ink droplets spread radially on paper. Dropped from a short distance, the image is a round splat. Dropped from a greater distance, the spot on the paper is a splash, with radial streams that look like fingers.

radially, in all directions. Second, the "natural selection" between the round splat and the splash with many fingers is a manifestation of the constructal law, and it is predictable [23]. The configuration that prevails is the one that offers the ink droplet greater (easier, faster) access to the surface on which it can spread and then come to rest.

The snowflake is another spreading flow that expands radially and displays a roundish form. What flows radially in the "solid" snowflake? Heat flows, by diffusion from the ice surfaces to the surrounding subcooled air. The source of heat is the latent heat released at the ice–air interface during solidification. In the beginning the ice is a tiny sphere. Later, the sphere grows needles in six directions, in a plane. Even later, three second-generation needles grow from the tip of each needle. The form of the snowflake becomes more complex as it matures, yet its broad outlook is roundish and dendritic.

The apparent diversity of snowflakes is similar to the diversity that we see in the ink spots of Figure 5.6. The natural choices between spherical ice, ice with six needles, and ice with two generations of needles are manifestations of the constructal law. The entire architecture (similarity, complexity, diversity) of snowflakes and other dendritic crystals is predictable [6, 24].

So much for the few shapes that appear in the geophysical and animal realms. It is fascinating that a few shapes have been emerging in the human realm as well, in the artifacts that have been empowering people during the recorded period. In the remainder of this chapter, we consider several examples.

First, why do boats with sails look the same? They have sails that are as tall (H) as the length of the hull (L); cf. Figure 5.7. They have hulls that are roughly ten times longer than they are wide (D_x). Furthermore, they are submerged to a depth (D_y) that is roughly the same as the hull width. The boat has been this way since antiquity (Figure 5.8). From a distance, the large boat looks like the small boat.

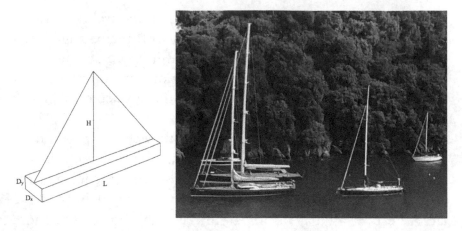

Figure 5.7   Boats with sail have the same shape. The height matches the length, and the width and depth of the hull are proportional to each other and to the hull length.

The reason for all these observations is the human tendency to move more easily on earth. The vehicle architecture that emerges is a reflection of the urge of all its builders and users to move more easily to have greater access to the surroundings. In recorded times, this tendency gave birth to artifacts (vehicles) in which people encapsulate themselves to acquire greater access. From this idea, the geometric similarity of all boats with sails is deducible in a few steps [25].

In the modern era, the physics principles of hydrodynamics [26] have played a guiding role in the improvements that have been made in the design of boats with sails. The icon of the central role of physics in boat design is Euler's entry [27] in the 1727 contest for the King's prize for the solution to the nautical problem to determine the best way to place the masts on vessels, and the relation between their positions and the number and height of the masts. After the arrival of steam power, fluid dynamics and naval engineering grew into distinct scientific domains, while naval vessel technology blossomed.

Airplanes are exhibiting their own "convergent evolution," and although mature, their evolution is much shorter than that

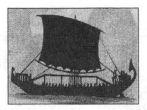

Time ⟶

Figure 5.8 Large boats with sails look the same. Top: From Egyptian galleys to Columbus and Napoleon's navy. Bottom: The Norwegian tall ship Sørlandet moored at Central Pier in Hong Kong.

of boats. A representative sample of models of commercial jet aircraft spanning five decades is shown in Figure 5.9. Airplanes are similar in their main aspects of geometry. The whole wingspan is the same as the fuselage length, the cross section of the fuselage is round, and the length/diameter ratio of the fuselage is roughly 10. The mass of the jet engines is approximately one-tenth of the total body mass. The fuel load and the range (the distance flown) are proportional to the body size. Furthermore, new models become more efficient from decade to decade: the specific fuel spent (for one seat and 100 km flown) decreased steadily. All these features have been predicted based on the physics of evolutionary design [28].

Helicopters, large and small, have converged on macroscopic features that unite them. The rotor diameter is the same as the body length. The proportionalities between body

Figure 5.9   Large or small, commercial jet airplanes look the same.

size, engine size, and fuel load are striking. The efficiency of the engine increases with its size. The specific fuel consumption has decreased steadily in the past five decades. These features unite the civil helicopters with the military helicopters, and they have all been predicted based on the physics of evolution [29].

Maritime ships with body sizes in the range of $10–10^6$ tons are united by similar macroscopic features and by the same physics of evolution [30]. The shapes and performance indicators of automobiles have evolved similarly. The geometric similarity of modern commercial aircraft, like the similarity of helicopters and automobiles, shows that human movement on the world map is facilitated by the generation and persistence of certain configurations in accordance with the physics of evolution.

The predictive power discussed above is very useful because it serves as the basis for fast-forwarding aircraft technology. For example, new construction materials and new jet engine designs may be adopted, but the movie on evolution physics has already revealed its script. The macroscopic features of the future designs that will prevail in time are natural, known, and worth implementing from the start of the new technology.

Nature continues to impress us with features and tendencies that repeat themselves innumerable times even though "similar observations" are not identical to each other. In science, we recognize each universal tendency as a distinct *phenomenon*. Over just two centuries, our predecessors have condensed each distinct phenomenon into its own law of physics, which now serves as a "first principle" in the edifice of science. A principle is a first principle when it cannot be deduced from other first principles.

The natural tendency toward organization in science is illustrated by the evolution of thermodynamics to its current state [31]. The transformation of potential energy into kinetic energy and the conservation of "caloric" were fused into one statement — the first law of thermodynamics — which has served as a first principle of physics since the mid-1800s. It was the same with another distinct tendency in nature: everything flows by itself from high to low. This phenomenon of one-way flow, or irreversibility, was summarized in another statement at the same time — the second law of thermodynamics — which serves as another first principle in physics. Now, the phenomenon of design occurrence and evolution with discernible direction in time — the constructal law — is a third, distinct phenomenon and self-standing law of physics (p. 2).

Why do the most common occurrences need such a long time to be recognized as natural tendencies (phenomena), and even longer to be recorded in physics with a short statement, a first principle? Because the evolution of the human mind — the subject of this book — is an integral part of the evolution of the human body and its movement (life). The evolution of the animal mind is to adapt and survive while struck by unexpected dangers — environmental, animal, and human. The first things that we tend to question are the unusual, the *surprise* (which, coming to English from French, means being grabbed from above, as if in the claws of a predator). We saw this in Chapter 2. The most common observations, the familiar and the non-threatening, are questioned the least. This is why *new* questions in science are very rare.

The few shapes that dominate the human realm circle back to the inanimate shapes with which this chapter began. Similar to the conical piles and plumes (Figures 5.1–5.3) are the piles of dry stone construction that dominated the ancient world. The pyramids of Egypt and Central America are icons of this human tendency to reshape the human niche, the landscape (Figure 5.10). The pyramid distinguishes itself through the

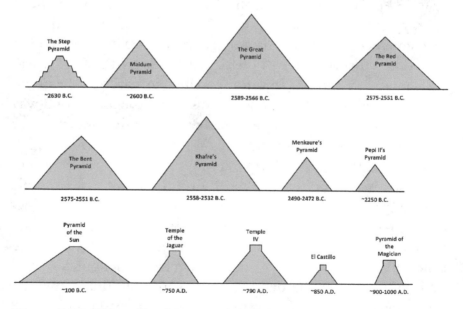

Figure 5.10   Pyramids of Egypt and Central America, chronologically, and their constructal principle of construction.

proportionality between its base and height, which is similar to that of the sand piles, smoke plumes, and boats with sails.

Like the sand pile, the pyramid grows one layer at a time while remaining geometrically similar to itself. It was shown [32] that this shape and its constancy during growth are manifestations of the human tendency to minimize the overall effort spent during construction. Even more, the angle formed by the sloped surface and the base is predictable from minimizing the work of bringing stones from the quarry on the plain to the top of the pyramid. The stones travel in two ways: easily, by sliding on the horizontal terrain, or with much greater difficulty as they are lifted one step at a time along the incline. Each stone follows a refracted path with a characteristic and predictable angle of refraction, much like the refracted ray of light and the refracted trade route on land or sea [33].

Paths form naturally where animals and people walk. On the side of a hill, the climbing path is snakelike, which is why it is called serpentine (from *serpens* for snake in Latin). It is a zigzag trail, where each segment is a grade (angle of climb) considerably smaller than the angle of the shortest path to the top of the hill. Going straight up is the difficult way, like lifting the stones step by step on the pyramid. Moving along the inclined segment is the easier way; yet, this way is also laborious when the angle of climb is small and, as a consequence, every segment of the zigzag is long, and many segments are necessary to reach the hilltop. The serpentine path itself, the number of turns and the length of each segment, emerges naturally from the animal urge to have easier access to the top. Like the shape of the pyramid, the serpentine path is the "shape" of the balance between difficult and easy, and between fast and slow.

In the human realm, the pyramidal shape is much older than the pyramids of Egypt. We know this from the Greek word "pyramid," which comes from *pyrá* for the pyre of wood made to light a fire. People all over the globe learned to shape fire the

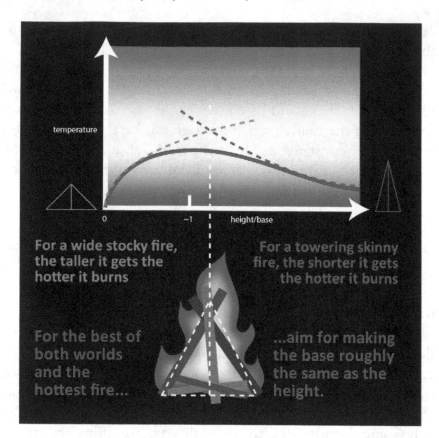

Figure 5.11   People all over the world shape fire the same way: the pyres are as tall as they are wide at the base. This shape makes the hottest fire per unit of fuel burned in still air.

same way: the height must be of the same size as the base [34]. The physics principle that governs this construction tendency is explained in Figure 5.11 [35]. When the pyre is too short, the buoyancy effect is minimal, and it is difficult for the surrounding air to be sucked into the fire to sustain the combustion. When the pyre is too tall, the surrounding air has easy access to the burning wood, but (because the entrained air is cold) it has the effect of cooling the fire. The trade-off between the short and tall shapes is the shape that everybody chooses unwittingly: the height is always comparable with the base width. This shape makes the hottest fire, and its significance is truly monumental. This very common shape of the fire represents the beginning of the evolution of energy technology, or the technology of power from fire, which has been essential for human evolution and empowerment.

Finally, we have the tree shape, which dominates and unites the animate and inanimate realms and is extremely common in human social organization. The movement of people and vehicles on earth is hierarchical because it flows through tree-shaped structures. Every movement is along tree-shaped paths, from the movement on city streets and in buildings to that in the global air and maritime traffic.

Everywhere we look, we see this evolutionary tendency. The footprints of a blue heron in the lower left of Figure 5.12 are

Figure 5.12 Tracks of a great blue heron in drying lakebed mud (Photograph: Garth Olson and Willem Larsen, with permission).

hard to distinguish from the fork-shaped cracks in the drying mud. The foot of a bird is shaped as a tree because this is the easiest way for its body weight to be transmitted as a flow of stresses to the horizontal surface of the ground. The "upside-down" bird foot is the design of the ceiling support in Terminal 5 at Heathrow Airport (Figure 5.13, left). There, too, the weight of the ceiling is collected by a small number of toes and then transmitted to the ground as a flow of stresses through the vertical bone.

As in the bird foot, all the "bones" of the tree support at Heathrow are in compression. The reverse, where the tree-shaped links are mainly in tension and bending, is much older, as on the right side of Figure 5.13. It is seen in how old wood beams flex to support the weight of the ceiling and what hangs from it.

The same tree design is the structure of all terrestrial animals. The palm, heel, and toes illustrate this. The lion tracks in Figure 5.14 are one example. The serpentine trail left by the beaver tail in Figure 5.14 draws the viewers' attention away

Figure 5.13   Inverted bird foot support for the ceiling of Terminal 5 at Heathrow Airport, and a chandelier pulling down the ceiling of the room in an old castle.

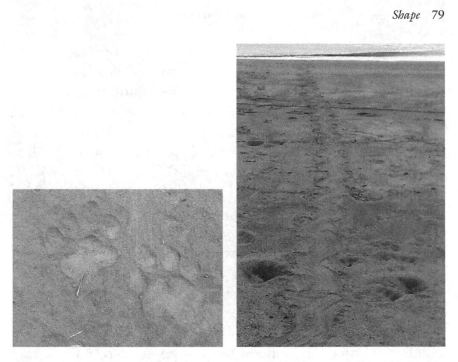

Figure 5.14 Left: Lion tracks in Lower Zambezi National Park, Zambia. Right: Beaver trail drag, going into the Columbia River (Photograph: Garth Olson and Willem Larsen, with permission).

from the footprints that alternate on both sides. The multitude of animal footprints (impala, cheetah, leopard) conveys two truths at the same time: tree design is predictable and diverse. It is like the "pedigree" of every being. It is not a coincidence that the word "pedigree" comes from the medieval French *pié de grue* and the Latin *pes + grus*, meaning crane's foot, just as in Figure 5.12. Disregard for the pedigree of ideas is damaging science, as we will see in Chapter 6.

The natural tree is arguably more beautiful than the human-made tree, such as the roof support in an airport (Figure 5.13). The reason is that the natural structure has been perfected over extended periods to serve more than one objective — that is, to facilitate more than one flow. In the botanical tree, the flow of water from the wet ground to the dry wind is ducted in the solid structure (wood) that facilitates the flow of stresses from the force of the wind to the ground, which holds the roots in place [36]. The stresses that flow with minimum strangulations endow the tree with its mechanical strength and resilience. So, this second objective (strength) is why every limb, small or large, is tapered toward its far end. It is tapered such that it is strong as it bends under its own weight.

Tapering is like perspective (Chapter 7), and unlike uniform thickness and parallel lines. Subliminally, the image that vanishes in the direction of the taper inspires the mind to see more than there is, making the image more attractive and memorable.

The tree shape is frequently celebrated in architecture. Two examples appear together in the same image (Figure 5.15): the live trees that decorate the plaza are seen through the constructal trees decorating the windows of a new engineering building at Duke. The window tree design came from Figure 1.3 in Ref. 6, which at the time of construction won the campus

Figure 5.15   Trees in the plaza and on the windows of a new building at Duke.

contest for the art that illustrates the oneness of life sciences and engineering.

To summarize the territory traveled in this chapter, let's remember that the occurrence of shapes is a universal phenomenon in all nature, animate and inanimate. At first sight, shapes strike us with their amazing diversity, which fuels our imagination and attraction to beauty and the sciences. In this chapter we learned that the diversity is underpinned by a surprisingly small number of shapes (cone, round tube, tree, boat, airplane). Each shape has its origin in the evolution of a *flow* architecture that is *free* to morph. The few shapes appear to be infinitely diverse because they morph all the time or because they had morphed for a long time in diverse environments, under time-varying circumstances.

The apparent contradiction between the untold diversity of shapes and their surprisingly small number is why the word *pattern* has emerged in descriptions of design in nature. To see the small number of distinct shapes is to see the pattern in the immense volume of diverse objects. The word *pattern* (from *patron* in French) has several related meanings having to do with a particular object considered worthy of imitation. There is pattern in the cloth that a tailor cuts along rigid templates on his bench, as there is similarity in the cookies that a baker makes with his cookie cutter.

The science of shape is about the physics of the evolutionary flow design that generates a small number of templates and cookie cutters. In the following chapters, we explore this physics under more familiar names such as message, perspective, art, science, and how to slow the speed of perceived time.

References

1. G. Tsatsaronis, Private communication, 21 August 2020.
2. J. S. Turner, *The Tinkerer's Accomplice*, Harvard University Press, Cambridge MA, 2007.

3. R. G. Kasimova, D. Tishink, Yu. V. Obnosov, G. M. Dlussky, F. B. Baksht, and A. R. Kacimov, Ant mound as an optimal shape in constructal design: solar irradiation and circadian brood/fungi-warming sorties, *J. Theor. Biol.*, **355**, 2014, pp. 21–32.

4. A. Bejan, *Advanced Engineering Thermodynamics*, 3rd ed., Wiley, Hoboken, 2006, Figure 13.21.

5. A. Bejan, *The Physics of Life: The Evolution of Everything*, St. Martin's Press, New York, 2016.

6. A. Bejan, *Shape and Structure, from Engineering to Nature*, Cambridge University Press, Cambridge, 2000.

7. A. Bejan and J. P. Zane, *Design in Nature*, Doubleday, New York, 2012.

8. A. Bejan and S. Lorente, The constructal law and the evolution of design in nature, *Phys. Life Rev.*, **8**, 2011, pp. 209–240.

9. T. Basak, The law of life: the bridge between physics and biology, *Phys. Life Rev.*, **8**, 2011, pp. 249–252.

10. A. H. Reis, Constructal theory: from engineering to physics, and how flow systems develop shape and structure, *Appl. Mech. Rev.*, **59**, 2006, pp. 269–282.

11. L. Chen, Progress in the study on constructal theory and its applications, *Sci. China, Ser. E: Technol. Sci.*, **55**(3), 2021, pp. 802–820.

12. L. Wang, Universality of design and its evolution, *Phys. Life Rev.*, **8**, 2011, pp. 257–258.

13. Y. Ventikos, The importance of the constructal framework in understanding and eventually replicating structure in tissue, *Phys. Life Rev.*, **8**, 2011, pp. 241–242.

14. A. Miguel, The emergence of design in pedestrian dynamics: locomotion, self-organization, walking paths and constructal law, *Phys. Life Rev.*, **10**, 2013, pp. 168–190.

15. R. G. Kasimova, D. Tishin, and A. R. Kacimov, Streets and pedestrian trajectories in an urban district: Bejan's constructal principle revisited, *Phys. A*, **410**, 2014, pp. 601–608.

16. L. E. Mavromatidis, A. Mavromatidi, and H. Lequay, The unbearable lightness of expertness or space creation in the "climate change" era: a theoretical extension of the "constructal law" for building and urban design, *City Cult. Soc.*, **5**, 2014, pp. 21–29.

17. G. Lorenzini and C. Biserni, The constructal law: from design in nature to social dynamics and wealth as physics, *Phys. Life Rev.*, **8**, 2011, pp. 259–260.

18. A. H. Reis, F. Miguel, and M. Aydin, Constructal theory of flow architecture of the lungs, *Med. Phys.*, **31**, 2004, pp. 1135–1140.
19. A. F. Miguel, Constructal pattern formation in stony corals, bacterial colonies and plant roots under different hydrodynamics conditions, *J. Theor. Biol.*, **242**, 2006, pp. 954–961.
20. A. Bejan, Why the bigger live longer and travel farther: animals, vehicles, rivers and the winds, *Nature Sci. Rep.*, **2**, 2012, 594, DOI: 10.1038/srep00594.
21. A. Bejan, Maxwell's demons everywhere: evolving design as the arrow of time, *Nature Sci. Rep.*, **4**, 2014, 4017, DOI: 10.1038/srep04017.
22. A. Bejan, S. Ziaei, and S. Lorente, Evolution: why all plumes and jets evolve to round cross sections, *Nature Sci. Rep.*, **4**, 2014, 4730, DOI: 10.1038/srep04730.
23. A. Bejan and D. Gobin, Constructal theory of droplet impact geometry, *Int. J. Heat Mass Transfer*, **49**, 2006, pp. 2412–2419.
24. A. Bejan, S. Lorente, B. S. Yilbas, and A. Z. Sahin, Why solidification has an S-shaped history, *Sci. Rep.*, **3**, 2013, 1711, DOI: 10.1038/srep01711.
25. A. Bejan, L. Ferber, and S. Lorente, Convergent evolution of boats with sails, *Sci. Rep.*, **10**(1), 2020, 2703.
26. E. Levi, *El agua según la ciencia*, MAPorrúa, Mexico, 2006.
27. Leonhard Euler's 1728 Essay to the French Royal Academy: E004, translated and annotated by Ian Bruce from *Meditationes super problemate nautico de implantatione malorum, quæ proxime accessere. Ad premium anno 1727 à Regia Scientiarum Academia promulgatum*, Paris, 1728.
28. A. Bejan, J. D. Charles, and S. Lorente, The evolution of airplanes, *J. Appl. Phys.*, **116**, 2014, 044901.
29. R. Chen, C. Y. Wen, S. Lorente, and A. Bejan, The evolution of helicopters, *J. Appl. Phys.*, **120**, 2016, article 014901.
30. A. Bejan, U. Gunes, and B. Sahin, The evolution of air and maritime transport, *Appl. Phys. Rev.*, **6**, 2019, article 021319.
31. A. Bejan, Evolution in thermodynamics, *Appl. Phys. Rev.*, **4**, 2017, 011305.
32. A. Bejan and S. Périn, Constructal theory of Egyptian pyramids and flow fossils in general, section 13.6 in Ref. 4.
33. A. Bejan, *Freedom and Evolution: Hierarchy in Nature, Society and Science*, Springer Nature, New York, 2020.

34. A. Bejan, Why humans build fires shaped the same way, *Sci. Rep.*, **5**, 2015, 11270, DOI: 10.1038/srep11270.
35. Duke University, The shape of a perfect fire, *EurekAlert*, 8 June 2015.
36. A. Bejan, S. Lorente, and J. Lee, Unifying constructal theory of tree roots, canopies and forests, *J. Theor. Biol.*, **254**(3), 2008, pp. 529–540.

6

Idea

An image conveys an idea to the viewer. Once grasped, the idea is out, like the proverbial cat out of the bag, or horse out of the barn. It cannot go back from the viewer's mind to the work of art on the page. Later, if the viewer reproduces the idea and presents it as his own, that is a stolen idea. My mother often warned me that "the eyes steal" (*oculi furantur* in Latin, *ochii furǎ* in Romanian), and this expression covers all the forms of the act, from copying during exams in school to peeping Toms and plagiarism in publishing.

If good, the idea keeps flowing on a growing territory from those who got it to those who would benefit from getting it. The growth is a one-way flow, just like the impression left in the eyes of the first viewer. The population reached by the idea, or the territory inhabited by that population, grows in an S-curve fashion [1], slow–fast–slow, as illustrated by the curves in Figure 6.1.

All growth phenomena exhibit the universal S-curve behavior because spreading consists of two mechanisms: first, spreading occurs along long and fast channels that cross the territory, and second, spreading is transversal to the channels, short and slow, from individual to individual. As shown in Figure 6.2, the first mechanism is channel flow, or "invasion," and the second is transversal diffusion, or "consolidation." Although the two

How ideas spread: S curves

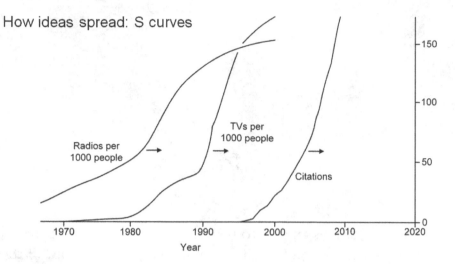

Figure 6.1   The universal S-curve growth of spreading over populations and territories.

ways to flow have a close connection from the beginning, the consolidation persists long after the invasion has spread across the territory.

During the invasion the size of the population (or territory) increases at an accelerating rate, and during the consolidation it increases at a decelerating rate. Spliced together, the two arcs form one S curve. The close association of invading channels (fingers, trees) embedded in areas (or groups) is the physics that unites all forms of S-shaped growth history, inanimate (river deltas, spilled milk), animate (plagues, animal body size; Figure 2.2), and social (ideas, news), spreading from point to area (or volume) and collecting from area (volume) to point. The universality of the phenomenon and the physics that underpins this universality are documented in Ref. [1].

The source of the original idea, which is the first viewing of the image, is the best instrument with which to establish whether the idea that reached you today is indeed original, because a picture is worth a thousand words. Finding the

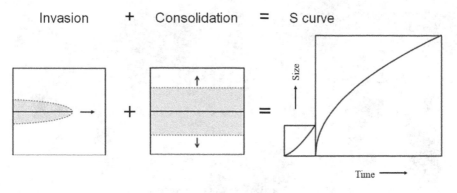

Figure 6.2 Spreading begins as "invasion" along channels (long, fast) and continues as "consolidation" by diffusion away from channels.

original image takes some effort, yet the good news is that in the digital era it has become a lot easier to search for it. The bad news is that it has become even easier to publish unoriginal ideas as "new."

I begin with a few examples of how to use the instrument proposed in the paragraph above. In each of Figures 6.3–6.13, I juxtaposed a recent image next to one that appeared before. I do not know whether the author of the new image was aware of the precedent, and I am certainly not accusing anybody of copying or committing plagiarism. I know even less if the older publication was the first of its kind. There are processes in place for identifying and addressing misconduct in scientific publishing (e.g., Ref. [2] and p. 100 in this book), and I will address them *in general* at the end of this chapter. Certain is that in every new case the precedent was not acknowledged. These figures are modest invitations to the viewer to compare, think, and question. Let the pictures do the talking.

Look at Figure 6.3, which is about world history, not science publishing. The 2009 image is the cover of the September edition of an academic magazine to which I subscribe. The upward, "bigger than life" look with the rising sun in the

Figure 6.3 The cover design of a bi-monthly magazine: a bolshevik propaganda poster, and the ever-present rising sun on the emblems of communist states.

background reminded me of a well-known image from another "brand new day" one hundred years ago. The rising sun in the background is also from one hundred years ago: it was on the emblem of almost every communist state. Evidently, propaganda imagery is forever effective, so much so that a graphic artist today uses that technique for great effect.

The next figures are from many examples in scientific publications [3]. I did not look for such examples: my readers keep sending them to me. The first theory that predicted the speed of an air bubble rising in a vertical tube filled with water was published in 1942 and 1943 [4, 5]. It was accompanied by a very good drawing (Figure 6.4, left), which became the icon of the phenomenon. The theory and the drawing were reproduced in 1967 [6] (Figure 6.4, right) without any reference to the precedent [4, 5]. One does not have to be an art critic to get the impression that in this particular case the new artist was looking at the original drawing while creating his version.

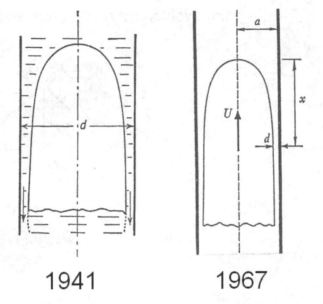

1941   **1967**

Figure 6.4 The air bubble rising in a glass tube filled with water: the original (left) and the remake (right).

Good images travel, and if we question the features of the image, we discover the source. The invasion of an area with flow channels, introduced here with reference to Figure 6.2, is illustrated in Figure 6.5 with tree-shaped channels that branch into crosses toward the periphery of a square area. In the original drawing, on the left of the figure, the stream enters from the lower-right corner of the square [7]. In the newer drawing, the stream enters through the center of the square. The two drawings may seem different, but that is questionable [8]. The original drawing is the same as one quarter of the new. It seems that the new drawing was constructed by repeating the original drawing four times.

Invasion over a square territory is illustrated in Figure 6.6 by the formation of palm-shaped channels. The channels grow by erosion of the substrate. In the original 1998 publication, the substrate is a porous medium through which water flows by

Area – point evolutionary flow architecture

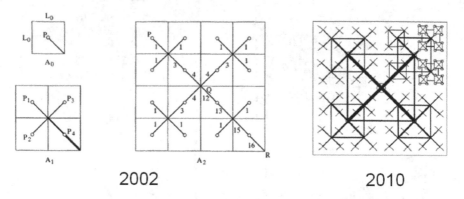

Figure 6.5   Tree-shaped channels bathing a square area: the original (left) and a newer version (right).

diffusion, called Darcy flow [9]. Grains are removed one by one when the water speed exceeds a critical value. On the right, we see a 2003 simulation of the formation of the same flow structure, except that the language is different: thermal diffusion (heat conduction) in a two-component material where the high-conductivity inserts are free to be moved around. The apparent similarities are discussed in greater detail in Ref. 10.

Spreading inside a volume is more complicated than over an area. It is in three dimensions, and it is more difficult to imagine and then draw. This flow phenomenon is very common, and so is its "invasion" architecture. It is called a vascular or dendritic system. Inhaling is one example of point-to-volume spreading, and the bronchial tree is its image. Exhaling is the volume-to-point version of the same phenomenon (i.e., inhaling in reverse), the same bronchial tree and image. The living tissue (muscle, organ) is a volume bathed by two of the previous trees, which are matched canopy to canopy. In the muscle, the arterial tree flows from point to volume and nourishes the volume. The venous tree flows from the volume to another point (outlet) and cleanses the volume.

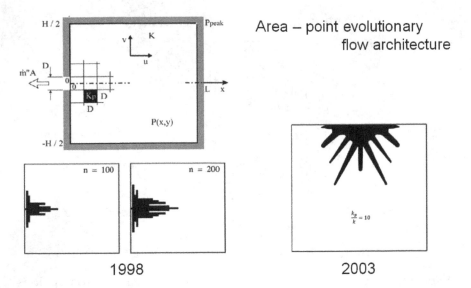

Figure 6.6 Simulations of the formation of a river basin on a square area discharging its flow to one point on the side: the original (left) was a model of water seepage through a porous medium; the newer version (right) replaced the water seepage terminology with a mathematically identical terminology (thermal diffusion).

In its simplest form, a volume is a three-dimensional box bathed over its entire space by one stream: inflow from one point to the volume and outflow from the volume to another point. This was the idea conveyed by the 2008 drawings [11] shown on the left of Figure 6.7. The first drawing shows how the single stream enters and sweeps the volume horizontally, vertically, and horizontally again (along the bottom), before exiting through the bottom-right corner. The boxlike volume is filled by two trees that share the same canopy, which is the volume itself. The second 2008 drawing (Figure 6.7, middle) shows how an even larger volume is filled in a tree–canopy–tree fashion by assembling (along the inflowing and outflowing shafts) a large number of elemental volumes of the kind shown on the left. On this background, the idea conveyed by the 2010

Point–volume–point vascular architecture

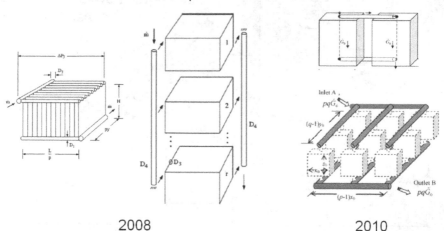

2008   2010

Figure 6.7   A finite volume is bathed by a single stream as two tree flows matched canopy to canopy. The stream enters from one point, sweeps the volume, and exits through another point. The volume that the two trees share is their joint canopy. Larger volumes are bathed in a tree–canopy–tree fashion by trees with more levels of branching, or confluence. The original is on the left, and the newer version is on the right.

drawing (Figure 6.7, right) appears to be the same as the idea conveyed by the old drawing [8].

To arrive at the point–volume–point double-tree design from principle is no small feat. This is the architecture that makes possible all the flows of the nervous system. Above all, the brain flow is the superposition of an immense number of point–volume–point flows, where the points of origin inhabit and define the volume, and each point of destination inhabits the same volume. Designers of artificial intelligence (AI) are arriving in modest steps (models, facsimiles) at the superposition of multiple point–volume–point double-tree architectures. These are naturally hierarchical and modular constructs of smaller constructs. They are eminently scalable, from smaller to larger systems, and from larger to smaller. Speaking of artificial intelligence, here is expert advice:

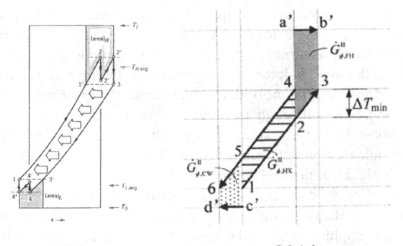

| 1988 | 2014 |

Figure 6.8 Counterflow heat exchanger between the hot and the cold ends of a power or refrigeration cycle: the original (left) and the newer version (right).

The hardest part isn't the coding. It's the *thinking*.

Costin-Andrei Oncescu
represented the University of Oxford
at the 2019 International Collegiate Programming Contest

An original 1988 drawing from thermodynamics (Figure 6.8, left) [12] provides a bird's-eye view of how the hot and the cold ends of an energy conversion machine (for power or refrigeration) are capable of functioning steadily at vastly different temperatures. This is because there is a long and highly effective counterflow heat exchanger between the two ends. The better the heat transfer between the two streams in counterflow (illustrated with horizontal fat arrows), the better the insulation effect in the vertical direction, between the top (hot) end and the bottom (cold) end of the power or refrigeration cycle. The drawing is original, and its idea is beautiful because it is counterintuitive: perfect thermal contact horizontally (stream to stream) inside the counterflow volume is synonymous with poor thermal contact vertically (along the streams), inside the same volume. The right side of Figure 6.8 is a 2014 drawing in a different notation that makes it appear new [13].

Heat flows through any space where the temperature is not uniform. Unlike a rotating shaft, which conveys the idea of macroscopic motion that can be seen and drawn, the flow of heat is invisible. Heat flow — the stream of energy that flows from hot to cold — is a concept, a mental construct that differs from one thinker to the next, and from one era to the next. Scientists who prefer graphics to formulas tend to indicate the flow of heat (called "heat transfer") with arrows — for example, the horizontal arrows on the left side of Figure 6.8.

When the medium through which heat flows is stationary, the direction of heat flow is perpendicular to the lines of constant temperature, called isotherms. This is called heat conduction, or thermal diffusion (p. 85). The lines followed by the flow of heat by conduction are called heat flux lines.

When the medium is in motion, the flow of heat is called convection, and its true direction is a lot more enigmatic. To draw the true direction of heat flow when the medium is

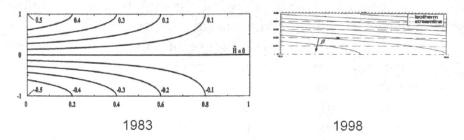

1983   1998

Figure 6.9   The paths of heat flow (heatlines) in a round tube with laminar flow (left), and a newer version of the idea, showing the isotherms and streamlines in the same flow region (right).

moving was the challenge faced by those who drew both sides of Figure 6.9 [14]. The two drawings represent the entrance to a round tube, with the hot fluid entering from the left. The tube axis is the midline of the left drawing and the bottom line of the rectangular image of the drawing on the right. In both drawings, the tube wall is the upper boundary, and it is cold.

The direction of heat flow by convection is due to two agents that drive the flow of energy through the volume. One is the temperature differences between isotherms, which in a stationary medium would drive the heat flux perpendicularly to the isotherms. The other is the flow direction of the medium, which in a tube flow (laminar, developed) is rectilinear and parallel to the wall (see the dashed streamlines on the right side of Figure 6.9).

The simultaneous presence of two agents is why the true direction of heat flow is visible in the original drawing (Figure 6.9, left). The heat current enters from the left, is swept downstream by the flowing fluid, and is gradually sucked by the cold wall. These lines have been known as "heatlines" since 1983, and if they land at equidistant points on the wall, they visualize the fact that the wall is heated (by the fluid) at a constant rate in the longitudinal direction.

On the right side of Figure 6.9, a more recent version shows isotherms and streamlines, not the flow of heat. To see the direction of heat flow, the viewer must execute two additional operations: one must draw the heat flux lines perpendicular to the isotherms and then draw the resultant of two directions (two arrows, if you will), the heat flux lines, and the streamlines. Needless to say, these two operations were already executed mathematically in the 1983 version, which yielded the formulas that were used in order to draw the heatlines precisely. The 1998 version of the idea looks different graphically because it stopped short of showing the resultant lines, the heatlines.

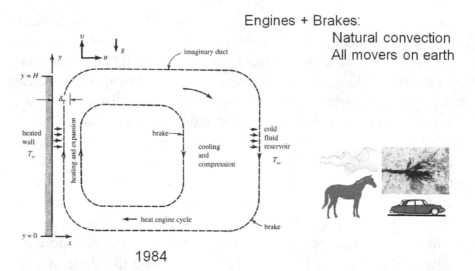

Figure 6.10 The heat engine and brake of the thermodynamics of all flows in nature: natural convection (left), and the atmosphere, hydrosphere, biosphere, and human sphere.

Convection occurs naturally in the presence of gravity. I showed in 1984 that any loop of "natural" convection (or "free" convection) is like a wheel driven by a heat engine [15] (Figure 6.10). The wheel of fluid rubs against the neighboring fluid, and the relative motion dissipates the power delivered to the wheel by the engine. All flow systems are governed by the same thermodynamics — winds, rivers, animals, and people (Figure 6.10, right). Each is an engine and brake system.

One of the most common examples of natural convection is the circuit executed by water in nature. This is shown in the 2008 tableau on the left side of Figure 6.11 [16]. The water wheel rotates counterclockwise, and the curtain of rain falls down and to the left. On the right side of the figure, the same features are present in a 2011 drawing [17]. Furthermore, the power cycle shown in the 2011 version is the same as the natural convection power cycle unveiled in the 1984 drawing (Figure 6.10, left).

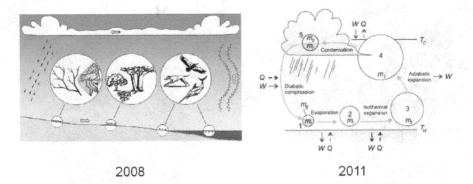

2008   2011

Figure 6.11 The circuit executed by water in nature (left), and a newer version of the same idea (right).

One of the ideas conveyed by the original drawing (Figure 6.11, left) is that the water flow along the earth's surface is facilitated by many evolutionary flow designs in addition to the most obvious — river basins and deltas. Less obvious are the flow designs of vegetation, animal locomotion, and human-made vehicles. The oneness of these moving designs is affirmed by the montage I made in the lower right of Figure 6.10 and on the left side of Figure 6.11. The tendency to have evolutionary designs unites plants with cars, rivers, and animals. All these designs have been evolving freely toward configurations that provide greater access and, in our era, have become "mature" [18] because they are not changing much after their long evolution.

Pairs of old and new are not restricted to scientific images (Figures 6.4–6.11; cf. Refs. 8, 10, 13, 14). Here are two examples from contemporary history. Figure 6.12 shows the Soviet jet fighter Sukhoi-27 (Flanker) next to the P. R. China version two decades later, called J-1 (Shenyang). The montage says it all. Analyses of this particular case are widely available, for example in a recent exposé by Roblin [19]. Stealing technology, scientific ideas, and credit for originality is rampant these days [20–37].

Figure 6.12   Two identical jet fighters: the original from the USSR, and the version from the People's Republic of China two decades later.

The second example is of two famous icons that are related in ways that most people are unaware of (Figure 6.13). On the left are the Olympic rings designed in 1913 by Pierre de Coubertin (founder of the modern Olympics) to represent the union of the Olympic athletic movement on the five continents of the globe. Yes, the five continents, and you can see them very clearly as three rings denoting the Northern Hemisphere and two rings depicting the Southern Hemisphere. The 1920 version, on the right, is the bolshevik red star, which was intended to herald the spread of communism on all five continents. The upper point of the red star represents the origin of communism, Europe.

As it turned out, in the race to conquer the five continents, the Olympic movement won. The irony is that policies enacted by all the communist states favored the success of the Olympic rings at the expense of the red star. Communist regimes barred their citizens from traveling abroad. Passports were forbidden because they would have enabled the people to run away after seeing the stark difference between home and freedom. At the same time, communist regimes funded an aggressive industry of "amateur" athletics designed to win medals at international competitions to prove the superiority of the communist system. Medals were won, but large numbers of athletes had to be allowed to travel to free countries to compete. These athletes

Figure 6.13 The graphic spreading of an idea on all five continents: the Olympic athletic movement versus communism.

became the flow of witnesses and television images that brought in every home the "difference" that the communist regime worked so hard to prevent.

Never mind the medals: sports are contributing the most to the spread of freedom, peace, hierarchy (the merit system), and understanding across the globe. The Olympics and the World Cup are at the top of this edifice envisioned by Coubertin. Along the way, the meaning of the words used in sports has a deep civilizing effect: athlete (from Greek *athlos*, a contest, and *athlon*, a prize) and amateur (from the Latin *amator*, lover, and *amare*, to love). Like children, we play and this way we become solid citizens [38].

Looking back at the examples of comparative art presented in Figures 6.3–6.13, my impression is that in each case the original is the better art, more beautiful and more inspiring than the remake. Art experts feel and react the same way: they rely on their first impression when they suspect a forgery and decide to investigate it.

It's the first impression that counts. The job candidate is advised to arrive well-dressed at the interview. The member of the jury knows from the first appearance the difference between guilty and not guilty. The first impression is usually the correct impression.

Original ideas are very few. As Frank Zappa remarked, "All the good music has already been written by people with wigs and stuff." It is much easier to take an existing idea, change a few lines in a figure and words in a text, and publish the same idea as new. The idea, not the exact figure or the exact text. This is why the National Science Foundation (U.S.) defines and enforces this kind of academic misconduct as follows [2]:

> Plagiarism means the appropriation of another person's ideas, processes, results or words without giving appropriate credit.

New name on a new publication does not always mean a new idea. On the other hand, a truly original idea deserves a new name, because otherwise the plagiarist will certainly name it after himself.

Plagiarism has victims. The word *plagiarius* means kidnapper in Latin, a person who gave his name to the child he stole. The Roman poet Martial coined the word when he became fed up with his poems being stolen. In modern times, a "plagiarist" means a literary thief. This meaning is sharply accurate. The mother recognizes her child's face in the crowd, and suffers. The victim of plagiarism recognizes his creation from oceans away, and suffers. Plagiarism is ugly.

> To work is too hard, and to steal is not beautiful
> (*Travailler c'est trop dur, et voler c'est pas beau*)
>
>   Zachary Richard

Plagiarism spreads like the plague because publishers and research-funding agencies are content to use software to detect pieces of text imported without credit from other sources. This is highly questionable. Science writing is not poetry and prose. In science one does not "copy," one steals the idea by looking. The eyes steal.

Even worse, publishers playact as enemies of plagiarism when they accuse the true author of "self-plagiarism." This is highly questionable. One does not steal from oneself. One owns what one creates. As I wrote recently [20], accusing the creative author of self-plagiarism is like accusing Picasso, Matisse, and Brancusi of thievery because they sold many pieces of art that looked like their own from a few years back.

One day, maybe, artificial intelligence (AI) will become intelligent enough to detect the unoriginal idea in a new publication. Until then, the detective work must be done by human eyes and brains.

I wrote a book about the emerging science of form [18], and in the present chapter I showed pictorially how to distinguish the remake of the idea from the original. This can be done by opening your eyes because pictures don't lie. Publishers, funding agencies, universities, and national academies must determine what misconduct is [20, 34]. Cheaters get away with cheating because administrators of our institutions (universities, journals) are not affected personally. Why? Because plagiarists do not steal from those who have published nothing worth stealing.

It is difficult to stop plagiarism when governments give monetary rewards to every author (and, indirectly, editor) who publishes one paper in a ranked journal [25]. It is difficult to stop plagiarism when the plagiarism experts invited on panels are the editors, not the victims [22, 23]. Most editors and publishers are unwitting enablers of plagiarism: new plagiarists cite old plagiarists in the same journal, the publisher and the editor reap benefits from the journal's rising impact factor.

Inspiration and theft happen because good ideas spread naturally [1]. Faced with this force of nature, scholarship has unwittingly arrived at a consensus: It is OK to get inspired, but you must acknowledge your source of inspiration. Cite what you read before you compose your version of it. Do this little thing, and you will be all right. Obviously, this little thing has not been taught uniformly across the globe. That is in pain view, cf. references at the end of this chapter.

To be healthy, an organization (university, government) must be rooted in individual conduct that is truthful and reliable. In corrupt academia, we can find hope in this Portuguese proverb: *A verdade é como a cortiça, vem sempre ao de cima* (Truth is like a cork, it always rises to the surface).

It does not require many words to speak the truth.

Chief Joseph

In closing this chapter, to the aspiring scientist I offer this connection with the preceding chapters: Time is change. Lifetime is a short movie of changes. In scientific work, "life" happens when you are busy contemplating new ideas. When new ideas stop occurring to you, then you are dead unless you are a plagiarist, in which case you prosper the way a parasite does. Parasite is not paradise. So, focus on change, new ideas, your ideas, your paradise.

To summarize, the big tableau illuminated by this chapter is populated by several images, each conveying a message to the viewer. The images are highly diverse, yet their messages have a few features in common. One is that messages are not equally important: a few travel far, while the great majority remain unnoticed. Another is that the important few are divided into two groups, the even fewer images that are original (the first of their kind) and the many remakes, forgeries, and stolen goods.

Friends and family members tend to tease me with jokes about the power to predict with the constructal law: past and future phenomena of evolution. The joke is more sinister than they think. The constructal-law power is so great that it predicts future "constructal" remakes that will continue to come out (we return to this in Chapter 10).

The fewest images are the most valuable and beautiful. In the next chapter, we learn how to make beautiful and realistic images, which is the legacy of the Renaissance artist who put "perspective" in the science of form.

References

1. A. Bejan, *The Physics of Life: The Evolution of Everything*, St. Martin's Press New York, 2016, chapters 7 and 8.
2. National Science Foundation, NSF-CFR-689.
3. A. Bejan, Nationalism and forgetfulness in the spreading of thermal sciences, *Int. J. Thermal Sci.*, **163**, 2021, 106802.
4. D. T. Dumitrescu, Strömung an einer Luftblase im senkrechten Rohr, *ZAMM — Z. Für Angew. Math. Mech.*, **23**(3), 1943, pp. 139–149, DOI: 10.1002/zamm.19430230303.
5. L. Prandtl, *Führer durch die Strömungslehre*, 1st ed., Vieweg und Sohn, Brunswick, 1934; 2nd ed., 1942; 3rd ed., 1949.
6. G. K. Batchelor, *An Introduction to Fluid Dynamics*, Cambridge University Press, Cambridge, UK, 1967.
7. S. Lorente, W. Wechsatol, and A. Bejan, Tree-shaped flow structures designed by minimizing path lengths, *Int. J. Heat Mass Transf.*, **45**(16), 2002, 3299–3312, DOI: 10.1016/S0017-9310(02)00051-0.
8. A. Bejan and S. Lorente, Letter to the Editor, *Chemical Engineering and Processing*, **56**, 2012, p. 34, DOI: 10.1016/j.cep.2012.02.007.
9. M. R. Errera and A. Bejan, Deterministic tree networks for river drainage basins, *Fractals*, **6**(3), 1998, pp. 245–261, DOI: 10.1142/S0218348X98000298.
10. A. Bejan, "Entransy," and its lack of content in physics, *J. Heat Transf.*, **136**, 2014, 055501, DOI: 10.1115/1.4026527.
11. S. Kim, S. Lorente, A. Bejan, W. Miller, and J. Morse, The emergence of vascular design in three dimensions, *J. Appl. Phys.*, **103**, 2008, 123511, DOI: 10.1063/1.2936919.
12. A. Bejan, *Advanced Engineering Thermodynamics*, Wiley, New York, 1988; 4th ed., 2016.
13. A. Bejan, Comment on "Application of entransy analysis in self-heat recuperation technology," *Ind. Eng. Chem. Res.*, **53**(47), 2014, pp. 18352–18353 DOI: 10.1021/ie5037512.
14. A. Bejan, Heatlines (1983) versus synergy (1998), *Int. J. Heat Mass Trans.*, **81**, 2015, pp. 654–658, DOI: 10.1016/j.ijheatmasstransfer.2014.10.056.

15. A. Bejan, *Convection Heat Transfer*, Wiley, New York, 1984; 4th ed., 2013.
16. A. Bejan, S. Lorente, and J. Lee, Unifying constructal theory of tree roots, canopies and forests, *J. Theor. Biol.*, **254**(3), 2008, pp. 529–540, DOI: 10.1016/j.jtbi.2008.06.026.
17. A. G. Konings, X. Feng, A. Molini, S. Manzoni, G. Vico, and A. Porporato, Thermodynamics of an idealized hyrologic cycle, *Water Resources Technol.*, **48**, 2012, W05527.
18. A. Bejan, *Freedom and Evolution: Hierarchy in Nature, Society and Science*, Springer Nature, Switzerland, 2020.
19. S. Roblin, Aircraft theft: Why China's J-11 fighter looks like Russia's Su-27 "Flanker," *The National Interest*, 19 December 2019.
20. A. Bejan, Plagiarism is not a victimless crime, *Prism. Am. Soc. Eng. Educ.*, **28**(7), 52 (2019).
21. How a Chinese student allegedly stole Duke University tech to create a billion-dollar empire, *today.com* [online]. Available: https://www.today.com/video/how-a-chinese-student-allegedly-stole-duke-university-tech-to-create-a-billion-dollar-empire-1284080195716. [Accessed: 30-Jul-2018.]
22. Science publishing: How to stop plagiarism, *Nature*, **481**, April 2012, pp. 21–23.
23. T. C. Long, M. Errami, A. C. George, Z. Sun, and H. R. Garner, Responding to possible plagiarism, *Science*, **323**(5919), June 2009, pp. 1293–1294.
24. T. Gabriel, Plagiarism lines blur for students in digital age, *The New York Times*, 8 January 2010.
25. W. Quan, B. Chen, and F. Shu, Publish or impoverish: An investigation of the monetary reward system of science in China (1999–2016), *Aslib J. Inf. Manag.*, **69**(5), 2017, pp. 486–502.
26. A. Abritis, A. McCook, and R. Watch, Cash incentives for papers go global, *Science*, **357**(6351), 2017, p. 541.
27. A. Qin, Fraud scandals sap China's dream of becoming a science superpower, *The New York Times*, 20 October 2017.
28. T. Mayer, Advice on dealing with research misconduct, 24 May 2016 [online]. Available: https://www.elsevier.com/editors-update/story/publishing-ethics/advice-on-dealing-with-research-misconduct. [Accessed: 30 July 2018.]

29. A. Bejan, Letter to the editor of renewable and sustainable energy reviews, *Renew. Sustain. Energy Rev.*, **53**, 2016, pp. 1636–1637, DOI: 10.1016/j.rser.2015.09.042

30. A. Bejan, Comment on "Study on the consistency between field synergy principle and entropy dissipation extremum principle," *Int. J. Heat Mass Transf.*, **120**, 2018, pp. 1187–1188, DOI: 10.1016/j. ijheatmasstransfer.2017.12.004.

31. A. Bejan, Letter to the Editor on "Temperature-heat diagram analysis method for heat recovery physical adsorption refrigeration cycle — Taking multi stage cycle as an example," *Int. J. Refrig.*, **90**, 2018, pp. 277–279, DOI: 10.1016/j.ijrefrig.2018.05.013.

32. A. Bejan, Evolution in thermodynamics, *Appl. Phys. Rev.*, **4**, 2017, article 011305 DOI: 10.1063/1.4978611

33. A. Bejan, Thermodynamics of heating, *Proc. R. Soc. A*, **475**, 2019, article 20180820 DOI: 10.1098/rspa.2018.0820

34. E. Chiscop-Head, Research integrity interview series: If you cheat, there should be a referee who blows the whistle against you, Duke University School of Medicine, 20 September 2019.

35. A. Bejan, Thermodynamics today, *Energy*, **160**, 2018, pp. 1208–1219, DOI: 10.1016/j.energy.2018.07.092

36. A. Bejan, Discipline in thermodynamics, *Energies*, **13**, 2020, p. 2487, DOI: 10.3390/en13102487.

37. A. Bejan and G. Tsatsaronis, Purpose in thermodynamics, *Energies*, **14**, 2021, p. 408. DOI: 10.3390/en14020408.

38. A Bejan, Watching physics at the Olympics, *Academia Letters*, 2021, article 2577.

7

Perspective

Nature is not math, formula, equation or algorithm. Nature is not made of two-dimensional objects, as in the images displayed in the preceding chapters. The objects of nature have four dimensions: the two dimensions of the image on the page, the third-dimension perpendicular to the page, and, as a fourth dimension, the discernible time direction in which the configuration of the object is changing. Nature is human perception and memory of movies on top of movies, all with the same plot.

The four dimensions are measurable with instruments, but are they all perceived by the mind? This is a fundamental question, because its answer applies to all the animals endowed with vision.

How the mind perceives the first two dimensions (the shape) was discussed in Chapter 3. How the fourth dimension (the time) is perceived was the subject of Chapter 2. What about the third dimension, the "depth" of the flat object printed on the page? Tough question, because we cannot dig into the page or photograph to discover the "thickness" of the object pictured as flat on the paper.

Animal evolution has been well ahead of such thinking and digging.

The animal mind was trained by the trials of life during the more than 500 million years that passed since the emergence of vision during the Cambrian explosion. The training boils down to a simple rule of understanding: things that are near look big, and things that are far look small. When the "things" are the same things, near and far, as in the images displayed in this chapter, the comparison is valid and the mental measurement of the difference between near and far is complete.

Perspective means to look through, in the direction of the depth, and it comes from the Latin *per* (through) and *specere* (to look). The word *perspective* captures what the animal mind does as it grasps the surroundings. The big and the small are compared, and when they happen to convey the same image (say, a lion, near or far), the animal understands the depth of the image and how close the danger is. On the contrary, if two different animals have the same size in the perceived image, the mind connects them both with similar images from the past, and the immediate impression is *similarity* (say, two quadrupeds, a lion and a dog) or *dissimilarity* (a wild predator versus a domesticated animal). The distance to the lion matters a lot more to the viewer than the distance to the dog even though they appear to have the same size.

The highway vanishes in the distance because it is long and far (Figure 7.1). The width of the pavement is the same everywhere, as can be seen in a photograph taken from above. The sides of the road are two parallel lines. When viewed in a direction close to the horizontal, two parallel lines and the base of the image form a triangle. The angle between the long lines becomes less sharp as the viewing direction gets even closer to the horizontal. The angle approaches zero when the view is from above, and as a consequence, the margins of the road appear parallel.

The animal mind did not acquire this training by looking at the painted pavement. Looking at the untouched surroundings

Figure 7.1   The converging road and clouds convey to the mind the third dimension of the image, the depth, or perspective.

was sufficient. The arrangement of clouds teaches how close the bad weather is, and from what direction it comes. In Figure 7.1, the clouds above the road are shaped as parallel bands. They are parallel rolls formed by warm, humid air rising (and its steam condensing) from under the visible clouds, and with cold, dry air descending into the gaps between the bands.

Why the cloud bands are parallel and equidistant is a thought worth contemplating while slowing down the passage of your mind time, as we will discuss in Chapter 9. Revealing is how Figure 7.1 conveys its depth. The spacing between the cloud bands decreases with "direction." Ones does not need to be in a car to understand that "ahead" means far.

By the way, animals too prefer the road because it makes walking a lot easier and safer than venturing through the bush. The dirt roads manicured for safaris are traveled by lions, elephants, and giraffes. Scavenging birds fly along highways to find roadkill, which has become a regular and dependable meal. Human movement influences animal movement, because both movements evolve in tandem to facilitate circulation, surface turnover, and climate on the globe.

Figure 7.2 The bare tree trunk serving as access route for squirrels, and the Stonecutters Bridge in Hong Kong.

A tall tree trunk offers another perspective and another highway. I see one right in front of my window (Figure 7.2, left). I became curious why squirrels climb fast on the trunk even though nothing threatens them on the ground. It finally dawned on me that for the squirrel the bare trunk is what the dirt road is for the lion. The squirrel gains altitude fast on the trunk, and then jumps laterally into the foliage of nearby trees, for food and safety.

The cable-stayed bridge pictured on the right side of Figure 7.2 reinforces the perspective of the tree trunk. Cables distribute uniformly the flow of stresses between the length of the roadbed and the height of the tower. The photograph was taken from a car as it approached the tower that stands on the centerline of the bridge, between the two bands of the two-way traffic.

Perspective, the sense of depth (distance), is perceived through other senses as well. All animals get it through hearing and touch (vibrations). The thunder announces the approaching storm with faint snapping sounds that become louder.

Simple as it may seem, the parallel lines that "converge" to convey the dimension of depth took a long time to be recognized in art. One can see this in how the technique of injecting perspective in paintings evolved from antiquity (Greece and Rome) to Renaissance, when Filippo Brunelleschi (1377–1446) formalized the rules of correct perspective drawing. Despite this, today the almost total dependence on machine graphics in scientific publishing produces images in which the parallel lines of nature are drawn parallel on paper, not convergent. This mistake is common, for example, on the right side of Figure 6.7.

Brunelleschi is a central figure of Renaissance architecture. He designed some of the most visited edifices in Florence, most notably the dome of the Florence cathedral Santa Maria del Fiore. He conceived and demonstrated the method of linear perspective in 1420.

The way to endow a drawing with depth (perspective) is illustrated in Figure 7.3. The lines in nature that correspond to the direction perpendicular to the paper must converge toward an imaginary point in the distance. If those lines are parallel, as on the actual object, the drawing does not look right on the page (see Figure 7.3, upper left). It gives the impression that the object becomes thicker in the distance.

The third dimension appears natural when the lines converge to a "vanishing point" far on the horizon. The cube drawn this way in the upper right corner of Figure 7.3 looks as if it resides on the left margin of the highway of Figure 7.1, in which the intersection of the converging lines of the highway is the definition of the vanishing point on the horizon.

When the cube is situated in front of the viewer, the natural look is captured with lines that converge to two vanishing

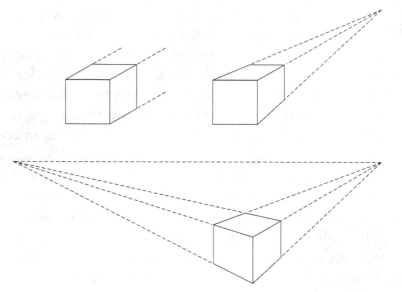

Figure 7.3 The difference between an easy-to-make drawing that is not realistic (upper left) and two more realistic drawings, with perspective. Depth is perceived when the lines that correspond to the third dimension (perpendicular to the paper) converge to one or two imaginary points in the distance.

points on the horizon, as in the lower part of Figure 7.3. This cube can be rotated to face the viewer by sliding the first vanishing point very far to the left. In this limit there is one vanishing point that remains, and the cube appears as in the upper right image of Figure 7.3. Finally, the remaining vanishing point must be moved to the left, to be in line with both the cube and the viewer, and the result will be the intended "three-dimensional" object perceived by the viewer.

The convergence of the "parallel lines" of nature is everywhere you look. It is so prevalent that it goes unnoticed. Just take a look at Figure 7.4, where the right wall of a hotel corridor looks like the left wall of a wine cellar. It is not necessary to spend a lot of time drinking in the wine cellar to have this

Figure 7.4 Symmetric corridor imagined by placing the wall of the wine cellar of Oriel College, Oxford, and the wall of the corridor in a hotel face to face.

illusion. The symmetric corridor is human nature, which is nature itself. In animal perception, parallel lines do not exist.

> The great artists are those who impose on humanity their special illusion.
>   (*Les grands artistes sont ceux qui imposent à l'humanité leur illusion particulière.*)

Guy de Maupassant

Illusion has more than one mother, and perspective is one of them. Tourists in Piazza dei Miracoli photograph themselves in positions giving the impression that they, with bare hands, are preventing the Tower of Pisa from falling to the ground. The same illusion is conveyed by the image on the left side of

Figure 7.5   Illusion from perspective: the big eats the small (left side), and the proper comparison between the two hippos (Photo: Teresa Bejan, with permission).

Figure 7.5. The image is believable because of Houdini's dictum (p. 6) and because of the common saying that the big fish eats the small fish. The truth about the two hippos is visible from a better angle, as on the right side of the figure.

By the way, the saying that the big eats the small is truer than it sounds. The big has the advantage and puts it to use because, in accordance with the constructal law, the big is faster than the small. This observation holds for all animal locomotion, swimmers, runners, and fliers. Even this generalization is truer than it sounds. Animals of the same size are faster in this sequence of the environments in which they move: water, land, and air. The speed of a flier is roughly ten times greater than the speed of a swimmer of the same size. The runner is in between — faster than the swimmer and slower than the flier. This is why in the food chain the big birds prey on fish and land creatures of the same size, while land creatures are capable of catching fish but must hide from big birds.

Figure 7.6 An image with perspective has the effect of liberating the viewer who is confined in a small room.

Perspective is liberating. Perspective is good for the mind because it gives the impression of "space" that in reality cannot be accessed. I took the photo shown in Figure 7.6 in a tiny (2.3 m × 2.3 m) waiting room with no windows, in a medical clinic. There were bare walls all around, except for this tableau in front of the chair on which I was sitting. Looking at and through the tableau I felt free, not incarcerated.

The perceived distance of "depth" matters even when the distance from the observed object to the observer is fixed in perpetuity. Think of the moon on the horizon versus how small it looks hours later, high in the sky. What matters in such cases is the size (length scale) in the details of the immediate surroundings of the observed object.

The mind perceives every new image comparatively, in relation to much older and more numerous images. The size of the full moon on the horizon is understood in relation to the tree-tops and rooftops that form the toothy contour of the horizon at dusk. Hours later, up in the sky, the moon looks a lot smaller because the object of reference (the sky) is much larger than the moon.

The same paradox is recorded in sayings such as being a big fish in a small pond, as opposed to being the same fish in the sea. The big fish from the small pond looks even bigger if it is the only fish in the small bag of a fisherman. In this direction of thought, the big fish in the small pond becomes a fish tale.

To summarize, perspective endows a two-dimensional viewing with a third dimension (depth) in achieving its purpose of conveying a message more clearly and faster. The addition of the dimension of depth to a scanned image that is made physically in two dimensions (made by the two eyes, Figure 3.3) was an important step in the evolution of art and science. We focus on this overarching phenomenon of evolution in the next chapter.

8

Art and science

The history of technology is recorded as a select parade of objects (devices, inventions, designs) aligned with the passing of time. The same can be said of the history of civilization, for which history books show us images of much bigger objects: edifices, roads, bridges, and aqueducts [1–5]. Understandably, in these books there is no room for the multitude of objects that preceded the commemorated few. Perhaps the omitted were not as good, or maybe their inventors were not as powerful to spread their use by trade, conquest, and writing.

The history of science is its own parade of images, from antiquity to the present [6, 7]. Yet, hidden under the feet of this marching column is the secret of the direction of the march. In the beginning, the objects of counting, arithmetic, and geometry were one-dimensional: lines, segments, size measurement, and the line axis of numbers. Later, the one-dimensional objects were joined by two-dimensional objects (plane geometry), and then by three-dimensional objects (solid-body geometry).

The direction has been toward liberating the form (drawing, design) to exhibit itself in more dimensions, and to convey its message more effectively. The direction has been toward greater freedom to change the configuration, from one dimension to two and three (Figure 8.1). More dimensions in designs mean greater complexity in the population of designs.

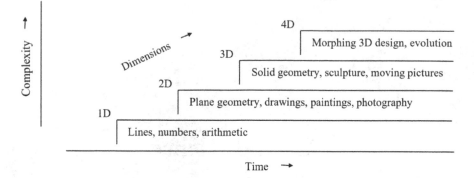

Figure 8.1   The evolution of objects toward more dimensions, complexity, and greater freedom to communicate their messages.

The same evolutionary direction is discernible in the history of art, from dots and line markings on cave walls to drawings on the same walls, all the way to the paintings and photographs of the modern era. This was the step from the one-dimensional to the two-dimensional (painting, photography, microscopy), after which both forms were retained. The one-dimensional was not discarded.

Next came the three-dimensional art, from bas-relief in ancient Middle East to Greek sculpture and, two hundred years ago, French descriptive geometry. One hundred years ago, a new kind of three-dimensional art arrived — the moving picture (cinema, animated drawings, cartoons) — in which the new dimension was time. This was the step from two dimensions (the plane image) to three dimensions (the plane image morphing in time). After this, the previous forms of object creation were retained [8–10].

The future of design will be four-dimensional: three-dimensional structures that morph freely in time. The fourth dimension is time, which is the measure of evolution and of what is "new" (the future). If the smile on my face is two-dimensional in a photograph, then my laughter is three-dimensional in a silent movie, four-dimensional in a movie with sound, and five-dimensional in a holograph with sound. In descriptive geometry, a three-dimensional object is created in space from two plane projections of the object, on the floor (as in "plan" view), and on the wall perpendicular to the floor (as in "profile").

The history of art and science makes obvious its time arrow, which points toward greater freedom for changing, morphing, and communicating the message from the observed object. This direction coincides with the path to performance, efficiency, economy, beauty, and access for human movement. For example, Figure 8.2 shows the evolution of human reliance on fossil fuels during the past two hundred years. Note the similarities between Figure 8.2 and Figure 8.1. The evolution has been stepwise. The new was added to the old. What worked was kept. The new and the old, combined, facilitated the movement (the life) of the user.

The history of human communications obeys the same time arrow. We think in terms of images, not symbols, lines of text, equations, and algorithms. The sequence from telegraph

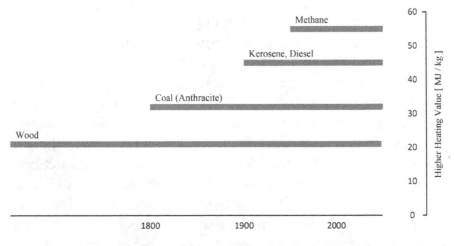

Figure 8.2 The evolution of the use of fossil fuels toward higher density of available energy, easier transportation, and greater access for users of the generated power.

to telephone and television is one exhibit in the overwhelming evidence. Most recently, when the internet era began, people posted all sorts of messages on the web. It did not take long for the world to respond with its own message: images are attracting many more followers than written text.

The evolution of communication is following the same constructal time arrow as in the Roman dictum *verba volant, scripta manent* — spoken words fly, writings remain. There is a saying that a picture is worth a thousand words. Now, a video is worth a thousand pictures and a million words. This is important to know because words are often being repeated to propagate falsehoods and lies. A single picture is more truthful than a thousand lies. In a court of law, visual evidence clinches the verdict. This is why visual imagery is so important (cf. Chapter 6).

Wind back the tape of communications to what was before electricity. The novelists who attracted the biggest numbers of readers were those who were very good narrators of "portraits" of chief personalities in their stories. The best historiographers were very good at painting in words the landscape filled with people building cities and defending them in battle.

Wind the tape back even farther. Language and counting emerged in base 10 because of the image of the two human hands. At preschool age we think 1, 2, 3... by hopping mentally from one finger on the next. We think in images. Even earlier, in human development, the counting was up to 3 because prehistoric humans knew the difference between going to the left, forward, and to the right. They knew the difference between past, present, and future, and between before, now, and next. This, the perception of time as change, was the start of the time arrow of mathematics, which passed (for our benefit) through the eras of arithmetic, geometry, algebra, and mathematical analysis. By the way, music rediscovered base 3 with waltz.

Today mathematics has become the most compressed (symbolic) language of science. As science progresses toward covering more and more aspects of nature, the mathematical representation of that convergence expands unabated. This is why many believe that a universal tendency (a phenomenon) must be a mathematical formula if it is to be a law of physics. From them we hear that the arrow of scientific progress is toward discovering the mathematical language of the universe. No, mathematics does not grow on trees and on undiscovered planets. Mathematics belongs to the individual who imagines it. It is an add-on to human language and rules of communication, and like all such add-ons, mathematics evolves to improve, accelerate, and compress human communication in synch with the swelling stream of science (knowledge) that deserves to be transmitted.

In all such examples, the time arrow that they share is no coincidence. It is the history of all designs (form, configuration) as one universal phenomenon of physics known succinctly as "evolution" [11]. It is no coincidence that the field of animal design based on principles of physics emerged in biology [12–17].

With evolution as physics, we are empowered to predict the future of objects, and this includes the future of design. If until now the history climbed on steps of one, two, and three dimensions, then in the future it will most certainly climb one more step, to four dimensions. The new generation of objects (art, science, design) that will join the earlier generations will bring objects that are not rigid — objects that morph to go with the times.

Continuing applications of evolutionary design (constructal law design) are demonstrating that the direction to greater performance is aligned with greater freedom to change the design. Examples are numerous and are catalogued regularly in review articles [18–20] and books [21–24]. A few designs are perfect, such as the round cross section of the duct for fluid flow, while many other designs are facsimiles of the perfect, such as the tree architectures discovered with minimal effort by minimizing the flow path lengths [24].

One generation after the constructal law field began (1996), some authors chose to repeat the same work by giving it a new name: topology optimization. First, topology and optimization do not belong together in English. Optimization means selecting from two or more *different* objects. Topology means the study of geometric forms that do not change even during deformation and stretching. The topology of triangles (linear or curvilinear) is called trigonometry. It is a discipline from 2,500 years ago. Authors who claim topology optimization are studying configurations that are changing and morphing constantly en route to selection; therefore, they are doing evolutionary (constructal) design.

Why do they do this? After all, calling it topography instead of topology would be closer to what goes on. The reason is that topography does not sound nearly as earth-shattering as topology. Similarly, one can easily rewrite his biography, but not his biology. One can move and change his geography, but one cannot change his geology — the discipline and the stones under his feet. And so, for want of easy fame and admission to academic clubs, one pastes holy Greek names (topology, biology, and geology) on his remake of older originals (to read more, see Chapter 6).

Perfect or not quite, a common feature of the superior flow architectures that have emerged along the evolutionary path of design is that each is one image — one rigid object — to be recorded in a new research paper, in a new production line in a factory, or on a new shelf in a shop. Even after all the freedom that the perfectly evolved design has had, it is now one rigid image.

Can the performance of the perfectly evolved object be increased further? Yes, it can be increased under those circumstances where the rigidity of the perfect form acts against its performance. This happens when the operating conditions of the object (the surroundings) vary in time. The perfect object that evolved with freedom in one environment is suddenly imperfect in an environment that has just changed. The object will perform below its ability unless it adapts to the new conditions.

The way out of this difficulty is once again freedom — more freedom to change the design to adapt to the times. The added degree of freedom is the ability to change in time. The object must have freedom to change as its environment changes. The object must be enabled to morph, to go with the times.

Before spreading on land, fish had evolved to the seemingly stable perfection that we see today in water. To move on land, fish needed changes in their undulating locomotion (one wavelength design). The change in the environment led to a longer undulating body (snakes, more than one wavelength), "walking" on the elbows of the S shape, and the emergence of bones and alternating legs under those elbows for terrestrial mammals later (amphibians, reptiles, and mammals) [25].

Thinking along this path, we predict nature. We do not describe nature. We do not copy nature.

One of the most common natural examples of this concept is shown in Figure 8.3. Grasses go with the times, and so do all the trees. They bend in the direction of the wind, and bend more when the wind is stronger. The usual explanation for why grasses and trees yield to the wind comes from biology, and it is expressed from a human-centered perspective: vegetation bends in order to "survive." This perspective predates science, as in this famous passage from *Antigone* [26]: "For a man, though he be wise, it is no shame to learn — learn many things, and not maintain his views too rigidly. You notice how by streams in

Figure 8.3 Vegetation morphs freely to go with the flow, and with the times.

winter time the trees that yield preserve their branches safely, but those that fight the tempest perish utterly."

The alternate explanation that comes from evolution (as physics) is that the vegetation architecture morphs in time to provide greater access to what flows. Inside the vegetation, water flows from the wet ground to the dry wind. The flow of water is facilitated when vegetation is not destroyed by strong winds. Outside the vegetation, air flows parallel to the earth's surface while overcoming the resistance posed by the ground, flat or with protuberances. The air flow is facilitated when the vegetation bends as if to get out of the wind's way. All the flows "survive" through their ability to morph in time.

Morphing to go with the times is a concept, an idea, a way to look with greater confidence into the future of human-made design. It is not, in any way, a blueprint for how to manufacture a particular design that morphs in synch with its changing surroundings. As in the movement facilitated by the evolution of fuel use (Figure 8.2), new industries and manufacturing techniques will surely evolve as the concept is understood, tried, and expanded in its applicability.

The evolution of technology is evident in the practice, devices, techniques, and concepts that empower us every day. It is evident in the evolution of science, research, and education. Old concepts, principles, and methods are replaced by more general, powerful, and concise new ones. The evolution is evident in how the design is changed and implemented. This is aided by steady improvements in simulation and computational techniques, optimization methods, genetic algorithms, printing, and manufacturing.

Form and design find themselves at a historic juncture.

The payback from the physics of mind time and beauty is that these two notions belong to the observer, the individual. This is the key — the mother of science — because science is not a few wild strawberries to be picked by a lucky wanderer. How I perceive time and beauty is not how you perceive time and beauty. This idea constitutes a new scientific development for three reasons:

(i) It corrects the physics book, which is a hundred percent the doctrine of the clock time, from nanoscales to cosmological scales, and which until now had nothing to say about the physics that underpins time, beauty, and other human perceptions.

(ii) It brings the observer into the physics book. This has been long overdue because all science is about us, which is precisely why science is useful to us, addictive, indispensable, and always insufficient.

(iii) All scientific research is autobiographical. It is about the mind, upbringing, languages, life, and times of the creative observer. The "autobiography" reaches a lot deeper than the opinion (choice, judgement) expressed by a scientist. One can change his opinion but not his flow of life.

This new development adds to the growing literature, showing that change (evolution) is physics and that biology, medicine, social organization, economics, and technology belong in physics. Engineering is physics because it covers the usefulness of physics, in the same way that medicine is the engineering of biology. Physics covers everything that is, the object and its immediate environment, the niche. The everything includes the observer and his or her niche.

The interesting science is about us and our lifetime. Every person and animal has the urge — the instinct — to live longer. Human life is movement that changes constantly. The movement is on areas, and it is like scanning. The easier movement is made possible by changes that occur freely in the configuration of the movement.

The physics of the mind time is particularly important today because many young people experience time distortion while spending too much time on social media. This has serious consequences, ranging from sleep deprivation to mood changes and mental disorder. It is why an unambiguous understanding of the physics basis of how humans perceive the passing of time is essential.

Science, the human contrivance kick-started as geometry and mechanics 2,500 years ago, continues to liberate and empower people to heights that were unimaginable even a few decades ago. Science has grown and spread spectacularly across the globe. It is improving hand in hand with the freedom to question, to emigrate, to make changes, and to evolve the current ideas, and designs.

We can't have enough of science [27]. Artificial intelligence (AI) is useful, but is it dangerous? There are some who think so; in fact, it is common to hear that AI is bound to kill and replace us. Well, tell that to those who need help, because good help is hard to find. They can't have enough of AI. In all the compartments of knowledge, people are contriving to make each of us

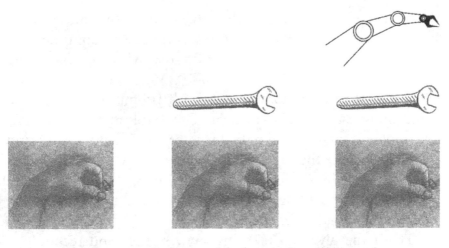

Time ———▶ Evolution of the human & machine species

Figure 8.4 The empowering of the human through the evolution of their "machine" part. From bare hand to hand with facsimile of hand (wrench), and hand with wrench and intelligent arm. Note that Figure 8.2 is another example of the evolution of the human and machine species.

a more powerful, longer-living, and faster-moving human and machine specimen (Figure 8.4).

Do not applaud the present too much if you do not know where the present came from. To primitive humans, counting occurred as 1, 2, 3, and later in base 10 (cf. p. 120). This quote from Wernher von Braun captures this truth:

"The best computer is a man, and it's the only one that can be mass-produced by unskilled labor."

History and the present are showing that science and technology are the great liberators. By adopting them we have more time to do other things. The urge to have more freedom to change and move things is natural and universal [11]. That is why, from the human point of view, technology is indispensable and addictive. We keep adopting new technologies, we keep doing "other things" with each chunk of the freed time, and we keep feeling that we have less and less time. And so, time flies, faster and faster in the minds of those who become more empowered (advanced).

Every new technology comes with promising impact and threatening impact. This is clear if we look at the past. The development of the steam locomotive, which put trains and railroads all over the globe, lifted the entire world out of poverty and, at the same time, took land from people who depended on land for food (grain, fruit, animals). Trains killed people, livestock, and wild animals.

That's the give and take that new science and technology offer, the promise and the threat.

Another example is nuclear power. It was developed during wartime, and today it is used peacefully all over the globe. Yet, nuclear power has dangerous aspects that keep us on alert and challenge us to be imaginative as problem solvers.

The internet offers great improvements in how people communicate, learn, and keep safe. At the same time, the internet lulls people (especially the young) into spending too much time in a virtual world, losing direct contact with real people and the passing of time. From this comes the alarming number of cases of mental illness. Furthermore, the internet is the new and much wider highway for fraud, from banking and human trafficking to fake news and false science in publishing.

Advances in science and technology are unstoppable because they come from individuals who create ideas that are useful to all. The moment of idea creation is a moment of intense intellectual happiness for the individual. I found that in the pursuit of happiness with ideas, ignorance is an asset. It is an asset because it does not dull the sharpness of the mental flash. The illusion of knowing too much breeds arrogance and foolishness. Respectful disregard for what other people would say is an advantage.

I owe the chance of making a discovery to my not being well-read.

Sigmund Freud

The dominant school of thought today is that before embarking on a new research project (a new idea, presumably), one must perform an exhaustive literature search to make sure that the idea has not been published already. We see this practice in every research proposal that we evaluate and in every research paper that gets published. The authors claim legitimacy by arguing that the literature shows that there is a "gap in knowledge," and that "there is a clear need to improve the understanding of...." This is false, because the literature says nothing of the sort. The one who says these things is always the author of the proposal and article, who selects and displays (and often omits, or misrepresents) the preceding work to bolster the claim that his idea is indeed new and needed desperately.

To see why ignorance is an asset, look at what happens if you — the idea creator — follow what is currently being taught. Follow the time axis in Figure 8.5. First, you are struck by a new

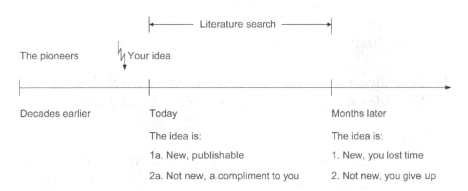

Figure 8.5 Two ways to act when struck by an idea: delay and death, because of extensive literature search, versus embracing the idea and giving it life right away.

idea today, and you spend months searching the literature. Never mind the effect of reading other people's work, which is a killer. You forget your own idea, or you become bored with it. The fact is that after this arduous search, you arrive at a fork in the road:

1. The idea was not published; therefore, you start your work on that idea many months after the idea found you.
2. The idea is already published; therefore you give up.

In both cases you lose.

The alternative is to pounce on your new idea as soon as it strikes you. Why? Because you were not born yesterday. You have kept your eyes and ears open to what your teachers have taught you. You are educated. After all, science is a story, and knowing the history of science is good for the soul. Because you are living in this story, you start working on your idea immediately.

After you do justice to your idea, you dress it up (or, better, you undress it and stretch it naked on the table) in a convincing form that gives you the courage to submit it for publication. This way you reach a different fork in the road, and you discover on which branch you are only after publication:

1a. Your idea is indeed original, and it is now published by you.
2a. Your idea was published by a pioneer, decades earlier.

In both cases you win.

If "2a" happens, it is a big compliment to you. It means that your thinking ranks up there with that of the pioneers who published something similar when people knew a lot less about anything. Your true peers are those pioneers. You are like them. Furthermore, it is extremely rare that "2a" happens, because an idea is a personal (intimate) reflection of its times, and it cannot be identical to what was conceived by another mind and written by another person in a different era.

In summary, by engaging in an extensive literature search (1, 2), the researcher and the idea lose time and identity. By pouncing on the idea (1a, 2a), the researcher, the idea, the science users, and science itself are the winners.

Idea-mining, like data mining, is an entirely different mental activity. There are many who search the literature permanently. They do not do that to verify the originality of their own ideas, because original ideas they do not have. They search to find old ideas that may be made to look new, if dressed in a different coat. It is much easier to change a few keywords and publish the idea as new. It is easier to acquire something through person-to-person interaction (contact, eavesdropping, watching, or theft) than to create it yourself. This technique is as old as school homework, public speaking, and publishing (see Chapter 6).

> Few discoveries are more irritating than those which expose the pedigree of ideas.
>
> Lord Acton

Those of us who still read the original sources know this, and when we get together, we marvel at how the good ideas (which travel far and persist; cf. constructal law, p. 2) are in fact not many, and are renamed along the way [28].

Do not look too hard for ideas: let them come to you, pace yourself, breathe, have rhythm. Sleep well at night, take naps in the afternoon. When ideas come, jump on them and you will be happy because they are you, and you belong in them. Do not plagiarize like so many do today, and do not impersonate authors who were better than you, and who are now dead and cannot protest. You are the ideas you create, and if they are good, they will outlive you.

The message of this chapter is this: Be happy in your ideas. Grasp them when they come, treat them with total respect, and share them with family, friends, colleagues, and the world. Research is not always easy. The proverbial snakes in the grass are everywhere.

> As in the sea between Scylla and Charybdis the helmsman is ever in danger, yet he will be thought shrewd and sagacious, if, keeping his ship on a straight course between the two, avoiding the rocks on the one side and the maelstrom on the other, he brings his ship safely to harbor:
>
> So in learning, the scholar is tossed between difficulties and adversities; but he will be worthy of praise and glory, if, directing his mind and proper reason around them, likewise avoiding any impediment or contention, he penetrates without hindrance into the Truth he seeks.
>
> Carlo Vitali (excerpt, translated from the *Dichiarazione dell'Impresa generale della nuova Accademia Peloritana detta de' Pericolanti*, Messina, 1729)

Here are three recent examples of how I follow my own advice:

First, the boundary layer is a classic concept in the contemporary physics of fluid flow. The idea is due to Ludwig Prandtl [29], who viewed (in his mind) that the fluid that flows past a solid body has two distinct regions, both invisible, a thin layer right next to the body surface, and the rest of the flow field. The thin layer was called boundary layer by Prandtl, and in it are located the shear stresses (fluid friction, dissipation of kinetic energy) that transfer the motion (momentum) from the fluid flow to the body surface, and vice versa (Figure 8.6). The rest of the flow space is the free stream, and dissipation is absent in it.

Incidentally, shear stresses and the dissipation of kinetic energy by fluid friction are invisible. They are mental viewings, and their sketch and narrative differ from one thinker to the next. One cannot photograph the dissipation that is distributed

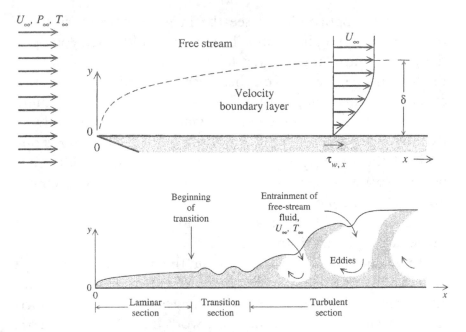

Figure 8.6 The invisible "boundary layer" on a plane wall parallel to a flowing fluid. The upper drawing is for laminar flow. The lower drawing shows how the laminar flow becomes turbulent sufficiently farther downstream.

through the flowing fluid. With imagination, however, you can visualize the dissipation. Spin a fresh egg on a smooth horizontal table. The rotational kinetic energy of the egg is dissipated by friction in two places where relative motion is present: externally, between the eggshell and the table, and internally, by the sloshing motion of the yellow sack inside the white sack. Dissipation is why the rotation of the fresh egg stops almost immediately. On the other hand, if you spin a boiled egg the rotation lasts a long time because the internal dissipation by fluid friction is absent, and the egg and the table are both smooth. The movie of the spinning fresh egg that comes to rest is the visual record of the effect of the invisible stresses and dissipation distributed through the fluid flow.

Isn't it fun how science empowers us to see the invisible?

Another opportunity to see the invisible is in every bathroom with running hot water. The water is cold when you open the faucet, and you must wait for the hot water to arrive from the heater. Imagine that your hands are not under the faucet to check the water temperature. You are busy doing something else as you wait. How do you know when the hot water arrived? By listening to the water jet. The viscosity of 70°C water is one third of the viscosity of 10°C water. The pressure difference between water supply and room air is fixed, and as a consequence, the change from cold stream to hot stream is accompanied by a noticeable increase in the flow rate and the noise of the jet as it splashes in the sink.

In science we indicate many of the invisible flows with arrows. The drawings can be memorable if they are beautiful and surprising. There are many montages of arrows — velocities, heat currents, mass flow rates, work transfer rates, electric currents, power and energy flows of all kinds, angular velocities, and angular momenta. This is where we see that science and art are one. They are about the same mental viewing — the idea. The arrows are as invisible as the flows. In the MIT cryogenics laboratory, where I grew up, we used to joke by walking "carefully" so that we would not be impaled by the arrows of heat transfer and angular momenta of the rotating machines all around us.

The mental image that an invisible boundary exists became the key to solving the fluid-flow equations that represented Newton's second law of motion. Prandtl's idea made it possible to predict the drag forces on airplanes. Those equations (called Navier-Stokes equations [28]) had been known for almost one hundred years, yet the wall friction problem proved impossible to solve, until the boundary layer idea. Boundary layer theory was a very big deal at the time, and because of it boundary layers were suddenly "seen" everywhere (Figure 8.7).

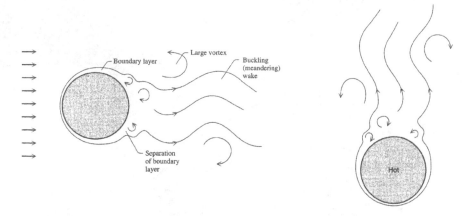

Figure 8.7 Boundary layers everywhere: On an object immersed in a stream (left), and on a hot object immersed in its own updraft (right).

I had no intention to write anything about an old idea as classical and accepted as the boundary layer, until an accident happened. I was composing a new homework problem [29] for my evolving (improving) course on convective flow and heat transfer [30]. The problem was about invoking the constructal law in the evolution of a flow configuration toward enhancing the spreading of movement (momentum transfer) away from a surface. After finding the solution and making a drawing of it, I was stunned: I had just predicted that the boundary layer and its shape *should* exist. I made this prediction in a completely different way, not ad hoc. I made it from a physics principle much more general than Prandtl's classical approach.

The problem and solution are sketched in Figure 8.8. Fluid is initially motionless in a rectangular space, the shape of which (H/L) is free to vary. The space $H \times L$ is neither slender nor tall. A blade enters at constant speed (V) along one of the sides of the rectangle and sets the fluid in motion by viscous diffusion. I reasoned that if the constructal law is valid, then the evolutionary direction should be toward the shape of the rectangular fluid space where the whole fluid is set in motion more easily,

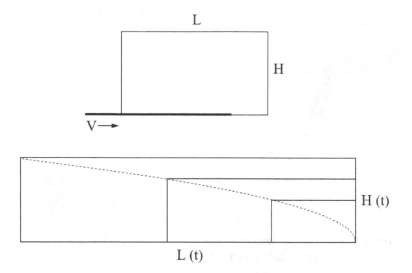

Figure 8.8 The shape of a fluid region set in motion completely during the shortest time. The shape becomes more slender when the size of the flow region increases in time.

during the shortest time. The shape is found with pencil and paper (with scale analysis), and it turns out to be the same as the shape of the classical boundary layer.

The difference between the two methods is fundamental. With the constructal law, the shape comes from the evolution of all the possible shapes toward greater access for the flow of momentum. With boundary layer theory, the shape is unique and comes from solving the simplified momentum equations for an assumed *slender* flow region.

The agreement of the two shapes (boundary layer vs. constructal) lends new physical meaning to why a boundary layer appears. It appears to set the surrounding fluid in motion more effectively, more completely, and faster.

Agreements between classical notions and constructal predictions have happened before. In the incipient rolling of slender flow regions (shear layers, jets, plumes), the natural selection of configuration (called "transition") between laminar flow and turbulent flow is always in favor of the configuration that offers faster access to the transversal transfer of longitudinal momentum to the neighboring layers. Predictions of the onset of turbulence cover a wide range of previously disconnected classical observations and correlations. For their derivation, terminology, and application, the reader is directed to my convective heat transfer textbook [30]:

— All the "critical" numbers of transition to turbulence (forced convection, natural convection, boundary layers, duct flow),
— The frequency of undulating (buckling) turbulent jets, wakes and plumes (the universal Strouhal number),
— The constant-angle shapes (wedge, cone) of turbulent flow regions with large-scale structure,
— The pulsating frequency of billows rising from pool fires,
— The smallest eddy that universally has a Reynolds number of order 10^2,
— The thicknesses of the viscous and conduction sublayers in fully turbulent flows,
— The analogy between fluid friction and heat transfer, and
— The analogy between rolling eddies and rolling stones.

In summary, turbulence is no longer an unsolved problem. Every step-size advance in scientific thought is a Gordian knot event. Once the knot is cut, the idea flow is liberated, and science and scientists move on. That happens, but not always. I keep noticing how the scientists, journals, and funding agencies that drew attention loudly to "unsolved problems" continue to draw attention to those problems after they have been solved. Turbulence is one example. Another is evolution, natural selection, and design in nature, where the discourse is still stuck in the search through the chance and randomness.

The knots are removed one by one by the laws of physics and, when necessary, by new laws of physics. And so, it is time to move on and ask new questions.

I end this first example with a true story from my first days of teaching convective heat transfer, which became the seed for my graduate-level course (1984), now in its fourth edition [30]. One of my colleagues, a young professor himself, asked me why I was making such a big deal out of the boundary layer and Prandtl. He said that the boundary layer is so obvious that somebody else would have discovered it soon after Prandtl. I told my colleague that he may be right except that if somebody else had thought of the idea, it would not be called "boundary layer" today. The words are Prandtl's, and they go with the image that had burst in Prandtl's mind.

I was the opposite of dogmatic about the boundary layer theory. Here is how I was teaching the idea and how I introduced it on the first page of its chapter in the 1984 edition of my convection course [29]:

1. No theory is perfect and forever, not even boundary layer theory.
2. It is legal and, indeed, desirable to question any accepted theory.
3. Any theory is better than no theory at all.
4. It is legal to propose a new theory or a new idea in place of any accepted theory.
5. Lack of immediate acceptance of a new theory does not mean that the new theory is not better.
6. It is crucial to persevere to prove the worth of a new theory.

Sadly, the most "confident" and "convinced" among us do not live long enough to wake up to the fact that they are being put to death because they had been wrong all along.

Moral certainty is always a sign of cultural inferiority. The more uncivilized the man, the surer he is that he knows precisely what is right and what is wrong. All human progress, even in morals, has been the work of men who have doubted the current moral values, not of men who have whooped them up and tried to enforce them. The truly civilized man is always skeptical and tolerant, in this field as in all others. His culture is based on "I am not too sure."

H. L. Mencken

The second accidental idea occurred at the start of the pandemic. I tell it here because it serves as "warm-up" to the next and penultimate chapter of this book. Today, many months after the start of the pandemic, scientists and journalists are asking why sub-Saharan Africa was relatively untouched by the coronavirus. *NBC News* put it bluntly: Covid models predicted devastation in Africa, but the reality is starkly different [31].

On March 3, 2020, at the onset of the pandemic, I predicted the African exception. A short article [32] is what I wrote then. The science that was published many months later [33] supports my prediction. Here are the main lines of the prediction:

(i) The nonuniform distribution of susceptibility to the coronavirus (Figure 8.9) corresponds to the nonuniform distribution of interbreeding with Neanderthals. After emerging through east Africa, the *homo sapiens* experienced three successive interbreeding periods that led to three types of newer *homo*, as summarized in Figure 8.10.

(ii) The human susceptibility was inherited from the Neanderthals, which is why those with zero Neanderthal DNA (in sub-Saharan Africa) are less susceptible.

(iii) The Neanderthals declined to extinction because of virus-induced diseases, not because of Darwinian "competition" with allegedly superior humans.

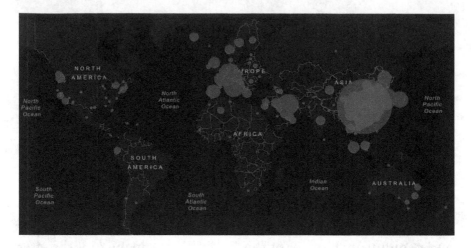

Figure 8.9   Global Covid-19 cases by CSSE, Johns Hopkins University, accessed 5 March 2020.

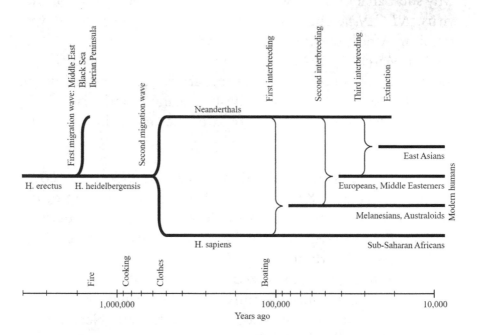

Figure 8.10   The spreading and mixing of humans on the globe [11]. The time scale is logarithmic.

A new mental connection is an opportunity for new scientific research. This means that researchers should take a fresh look at the available body of data on virus spreading and ask new questions about known facts.

The new understanding makes it easier to anticipate the spread of diseases and to offer help to those who are threatened the most. In retrospect, it was fortuitous that I made this prediction very early during the pandemic, before the whole world became infected, and before governments intervened to control the spread. Today it would be impossible to see the connection that the early spreading revealed (Figure 8.9).

It was fortuitous to connect the virus susceptibility to the chronology and geography of the interbreeding with the Neanderthals, because this shed a new light on the extinction of the Neanderthals themselves, in addition to the spread of future pandemics.

Incidentally, in science, it is essential to keep an eye on the geography of the object. Everything has "geography," because geography is the graphic description of movement on purposeful paths on earth. This is true of all animal life, human life (thought, science, city, economics, trade, science, government, war), the winds, the rivers, the ocean currents, and the climate. If we pay attention, everything that matters has geography, which is why geography too is physics.

The third human event that sparked an idea took place across the street from my office. I was invited to attend Prof. Ehsan Samei's research group meeting in the medical school. The effort that unites the group is directed toward imaging the growth of cancerous tumors. We talked freely, students and professors, and at the end of the two hours, a new point of view was emerging on the blackboard: how to predict the flow architecture that would later be imaged in radiology and tracked as a "tumor." The theory was just published [34], and it attracts attention.

The seed inside the core of the idea is the tendency of all flow systems to morph in freedom to provide greater access to their currents. Evolutionary designs and growing systems are manifestations of this tendency. In the present example, cells come together into clusters that grow, reorganize and vascularize in order to create greater access to flowing nutrients. Unlike in traditional approaches that focused on the cell and cell clusters, our view was holistic: the system (our drawing) was the cell cluster embedded in its immediate surroundings, which is called extracellular matrix (ECM).

This new kind of system — cell and ECM — turned out to be the key to being able to predict the stepwise growth of both, cell cluster and ECM (Figure 8.11). We predicted the one-to-one relation between cluster size and structure (configuration), and the critical cluster sizes that mark the transitions from one

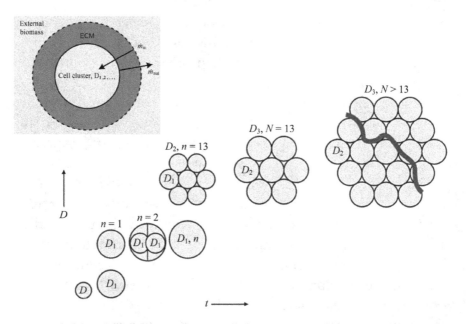

Figure 8.11   Cell cluster and extracellular matrix, and the stepwise changes in the flow architecture as the size of the cluster and its matrix increases [34].

distinct configuration to the next. This kind of thinking teaches the research student what theory is, and why it is rare and very useful. Theory explains not only how growth should occur, but also why it must occur. Theory is the crystal ball.

In summary, the lesson from this chapter is simple: When you work on an idea for too long, stop. Start working on a new idea. Return to the first idea later. Kick more than one ball. This is a good way to slow down the mind time, as we will see in the next chapter.

References

1. F. Klemm, *A History of Western Technology*, The MIT Press, Cambridge, MA, 1964.
2. A. F. Burstall, *A History of Mechanical Engineering*, The MIT Press, Cambridge, MA, 1965.
3. L. Sprague de Camp, *The Ancient Engineers*, Barnes & Noble Books, New York, 1993.
4. A. M. Liberati and F. Bourbon, *Ancient Rome*, Barnes & Noble, New York, 2006.
5. M. Swift, *World Cities Yesterday and Today*, Fall Rive Press, New York, 2009.
6. A. C. Crombie, *The History of Science from Augustine to Galileo*, Dover, Mineola, NY, 1995.
7. B. Atalay, *Math and the Mona Lisa*, HarperCollins, New York, 2006.
8. A. Seba, *Cabinet of Natural Curiosities*, edited by I. Müsch, R. Willmann and J. Rust, Taschen, Hong Kong, 2005.
9. E. Haeckel, *Art Forms in Nature*, Prestel, Munich, 2008.
10. C. Keller, ed., *Brought to Light: Photography of the Invisible, 1840–1900*, San Francisco Museum of Art, 2008.
11. A. Bejan, *Freedom and Evolution: Hierarchy in Nature, Society and Science*, Springer Nature, Switzerland, 2020.
12. K. Schmidt-Nielsen, *Scaling: Why Is Animal Size So Important?* Cambridge University Press, Cambridge, UK, 1984.
13. S. Vogel, *Life's Devices: The Physical World of Animals and Plants*, Princeton University Press, Princeton, NJ, 1988.

14. E. R.Weibel, C. R. Taylor, and L. Bolis, *Principles of Animal Design*, Cambridge University Press, Cambridge, UK, 1998.

15. B. Ahlborn, *Zoological Physics*, Springer, Berlin, 2004.

16. A. Bejan, *The Physics of Life: The Evolution of Everything*, St. Martin's Press, New York, 2016.

17. A. Bejan and J. P. Zane, *Design in Nature*, Doubleday, New York, 2012.

18. A. H. Reis, Constructal theory: From engineering to physics, and how flow systems develop shape and structure, *Appl. Mech. Rev.*, **59**, 2006, pp. 269–282.

19. L. Chen, Progress in study on constructal theory and its applications, *Sci. China, Ser. E: Technol. Sci.*, **55**(3), 2012, pp. 802–820.

20. L. Chen, H. Feng, Z. Xie, and F. Sun, Progress of constructal theory in China over the past decade, *Int. J. Heat Mass Transf.*, **130**, 2019, pp. 393–419.

21. L. A. O. Rocha, *Convection in Channels and Porous Media: Analysis, Optimization, and Constructal Design*, VDM Verlag, Saarbrucken, 2009.

22. G. Lorenzini, S. Moretti and A. Conti, *Fin Shape Optimization Using Bejan's Constructal Theory*, Morgan & Claypool Publishers, San Francisco, 2011.

23. A. F. Miguel and L. A. O. Rocha, *Tree-Shaped Fluid Flow and Heat Transfer*, Springer Nature, Switzerland, 2018.

24. A. Bejan and S. Lorente, *Design with Constructal Theory*, Wiley, Hoboken, 2008.

25. A. Bejan, The constructal-law origin of the wheel, size, and skeleton in animal design, *Am. J. Phys.*, **78**(7), 2010, pp. 692–699.

26. Sophocles I, *Antigone*, 2nd ed., translated by D. Greene, University of Chicago Press, Chicago, 1991, p. 690.

27. A. Bejan, AI and freedom for evolution in energy science, *Energy AI*, **1**, 2020, 100001.

28. A. Bejan, Nationalism and forgetfulness in the spreading of thermal sciences, *Int. J. Therm. Sci.*, **163**, 2021, 106802.

29. A. Bejan, Boundary layers from constructal law, *Int. Commun. Heat Mass Transf.*, **117**, 2020, 104672.

30. A. Bejan, *Convection Heat Transfer*, 4th ed., Wiley, Hoboken, 2013; 1st edition 1984.

31. D. Chow, Covid models predicted devastation in Africa — but "the reality is starkly different," *NBC News*, 25 September 2020, https://news.yahoo.com/africa-held-off-worst-coronavirus-100013035.html.

32. A. Bejan, Coronavirus invasion and Neanderthal retreat, *The Next Truth*, **3**(4), February 2021, pp. 11–14.

33. H. Zeber and S. Pääbo, The major genetic risk factor for severe COVID-19 is inherited from Neanderthals, *Nature* (2020). https://doi.org/10.1038/s41586-020-2818-3.

34. T. J. Sauer, E. Samei, and A. Bejan, Cell and extracellular matrix growth theory and its implications in tumorigenesis, *BioSystems*, **201**, 2021, 104331.

9

Slowing time

In everyone's life, the mind time began to speed up early. Admit it, when you were very little you thought that the clouds were motionless, just hanging there, as part of the sky. They seemed motionless because your tiny eyes were jiggling at high frequency. Time stood still then, which is why old people say that time is being wasted on the young. Then came a moment when you noticed for the first time that the clouds were moving. If you were surprised by your discovery, then you remember that event today. If you remained curious about the clouds and what their shapes teach you, then you noticed that the clouds have been moving faster as you got older.

Slow time becomes fast time during every activity that repeats itself the same way many times. You noticed this as a student. The first lecture in a course is slow, new, and memorable, and so is the second. The trailing lectures are considerably less memorable. Here is another example: imagine that a family member just had eye surgery, and your role is to administer six drops at precise times every day, for four weeks. That's a lot of eye drops and a lot of attention that you must pay to the patient and the clock. In the beginning, it feels like a full-time job that will never end. Yet, in the third week you are wondering where the time went.

It's never like the first time. This impression is captured perfectly by the title of the 1962 film *The Longest Day*, which is about the Normandy landings (D-Day).

When the lockdown began in April 2020, I felt that time slowed down. Many of us felt it, and talked about it on the web. Right away, news commentaries started to ask why this was happening. It did not take long for journalists to ask me to explain it in view of the physics theory of one year earlier [1], which had been in the news constantly (Chapter 2). Journalists and correspondents continue to ask, so here is a synopsis of that discussion, thanks to the interesting writers with whom I spoke [2–15]. It is an important discussion to have because if we know why the time slowed down for so many of us, we also know how to slow it down for our own well-being in the future.

During the first days of the lockdown, we were forced to experience new things. The new lifestyle resembled the life that existed before the industrial age. We stayed close to home, and we moved slowly. Those of us who were living the routine of the fast lane (job, automobile, airplane) were suddenly forced to quit. We were forced to move less. As a consequence, we had time to listen and talk and, surprise, our sleep improved, and our dreams became deeper and easier to remember.

To keep active, I walked every afternoon in the park near our home. I noticed that the park had become full of people, my neighbors. We greeted each other. A week later, the neighborhood was connected more tightly than before. This sudden change was not new. I noticed it during Hurricane Fran (September 1996), which had toppled many of the trees, blocked the traffic, and interrupted the electric power for a whole week. Neighborhood life had come to a standstill then, and it reminded me of life in the peasant society with which I was familiar in my childhood.

In the park every afternoon I was doing many new things. I was observing, seeing, and hearing. The other walkers were doing the same. I was looking at everything — trees, flowers, birds, insects, people with their dogs, and the ground on which I was stepping.

This new life did not last long. As the lockdown continued, so did the walks, the talking, and the listening. Two weeks later, the new life was not new anymore. It had become its own routine. In the park, people stopped noticing each other because they were talking on their iPhones. Walking was their way of working from their socially distanced offices. Six months later, they do not even look at me when I wave at them.

Quietly, without warning, time had regained its faster pace. The connection between routine and fast time, or between new experiences and slow time, could not have been more evident. As I show below, this is how I discovered how to predict what will happen next, when the lockdown ends along with the new form of house arrest called "teaching online."

By the way, the image of house arrest reminds me of my father's favorite metaphor of protest during the most brutal period of the communist regime, the dictatorship of the proletariat in the '50s and '60s. He was a veterinarian, and because dogs in the Romanian village are on a chain guarding the yard, he felt safe enough to say loudly: Look in the eyes of the dog; he is telling you, "I want to be free." A former classmate from the same place and era recently sent me the reverse of that metaphor: Under the lockdown, I am like a dog. I wear a muzzle, I have a certificate that I have been vaccinated, and I need permission to leave my yard.

When we go back to our regular jobs and staff meetings, the mind time will slow down again, at least for a while. It will feel good because those meetings will be new experiences. Interacting with people gets the mind going. Yet, good news does not last.

The mind time will speed up again when the pre-lockdown routine returns.

I tested this prediction on myself, and I found that it is correct. Two weeks after the lockdown, I interrupted my new routine by going to my office to videotape my lectures for my students online. I did this with chalk on the blackboard in my laboratory. I did it twice a week, and it felt good. Time slowed down.

Why teach with chalk on the blackboard? Because the computer (the "online" experience) is not academia. It does not come even close. True academia, the classroom, is a physical space in which teachers and disciples interact and teach each other, in counterflow. Contrary to that, online teaching is a new type of top-down indoctrination. The teacher preaches, and the students bow, in silence.

The live audience during the filming of my lectures was the filmmaker himself, Sinan Gucluer, a visiting professor from Aydin Adnan Menderes University in Turkey. My audience was real, not virtual. Gucluer would interrupt the filming to ask questions and tell me how to teach better. Now my lectures are more permanent because they are available on film for audiences that I will address in the future. This is a significant advancement because my lectures were already available in print, in my textbooks. How they became available in print is the result of a slowdown very similar to the one of April 2020. Here is that story, as I wrote it in the preface to the fourth edition of my convection treatise [16]:

At age 33, I was behaving as if I was meant to play basketball forever, and I was wrong. During a game in January 1982, one of my Achilles' tendons ruptured and I ended up in a wheelchair for the entire semester. I had to teach my convection course for which I had written notes, but this time I was forced to write every lecture on transparencies for overhead projection on the screen in the classroom. My first doctoral student, Shigeo

Kimura, now vice president of Komatsu University, Japan, was my teaching assistant. He would wheel me into the classroom every morning, and my new convection book would come to life, one original drawing at a time, one original (solved) problem after another. There was so much time and creativity during the spring of 1982, all from an accident.

Change is key. Change makes an impression. Change revs up the creative mind. The imagined "next" may turn out to be a mirage, yet to imagine "change" is always healthy because it nurtures hope. The urge to imagine change is captured in the saying that the grass is greener on the other side of the river.

Growing up, I acquired the habit of looking at animals, paying attention to their looks and behavior. You can do this yourself for something new, to rescue yourself out of the stupor of your routine. Install a bird feeder close enough so that the birds can watch you. If you do not move at all, the tiniest birds keep feeding as if you are absent. As soon as you twitch — cheek, finger, or shoulder — they fly away. Slightly larger birds (cardinals, woodpeckers) fly away sooner because they recognize the bigger change in their newly perceived image: your silhouette.

From infancy, the brain is accustomed to receiving new impressions during finite time intervals. The expectation is to receive a certain number of impressions while awake during daylight. When that number increases, the day feels longer (time slows down), as during the start of a lockdown and other unexpected changes. When that number decreases, the day feels shorter (time accelerates), as during any routine, past or future. The brain has been trained this way long before speech, writing, and clocks empowered the individual.

Change is how to slow down your mind time. During the silence that fell when the lockdown was imposed, I used my own theory of how to slow down my mind time. While walking, I looked more carefully at images that otherwise would have gone unnoticed. After all, how often do you count the ants on which you are stepping? You don't count, because you don't even think to look. On the other hand, if you look, you may think before you step on life.

The new impression that comes from looking at the unsuspected is a new mental connection. It is a new mental viewing that recommends itself to be recorded because it fits among similar impressions from the past. It is "recognized" because past impressions exist in your memory. If your beloved dog of 20 years dies suddenly, grieve, and then get a new dog that looks like the old one, and give him the same name.

The solidity of the memory "tetris" wall is good for continuity in life. The new, on top of the old, empowers the mind to imagine the future. We saw this aspect of design in how the mind understands perspective (the depth of the two-dimensional image) from the difference between the sizes of similar objects (cf. Chapter 7).

Memory is how I discovered the face of the moon in front of my feet, in the park. The color and craters of the moon are obvious (Figure 9.1, left). I was astonished when I saw this, and I will not forget it. The new mental connection is between the real moon and the craters that appeared suddenly on the sandy

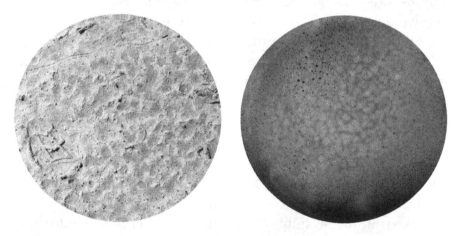

Figure 9.1   Moon craters and sunspots at your fingertips, if you keep your eyes open.

path in front of me when the summer rain started to impact the ground with big and heavy drops (Homework for the student: predict that big raindrops fall from high altitude, and small raindrops fall from near the ground).

The surface of the sun appears in your espresso before you take the first sip. The sunspots are clearly visible on the right side of Figure 9.1. This is no coincidence. The image is caused by the same kind of subsurface flow as on the sun. It is called Bénard convection [16], or natural convection in a fluid layer heated from below. We cannot see the flow under the surface (sun or coffee), but on the surface we can see the fingerprint left by the buoyant flow. Many plumes of hot fluid rise, reach the surface, cool off, and sink into the spaces between the rising plumes. On the coffee surface, the tops of the rising plumes are darker (like the pure coffee from below), and the sinking spots are lighter because of the foam that gathers above them. On the solar surface, the tops of the rising plumes are brighter (hotter), and the sinking spots between them are darker (colder).

The same tapestry of dark and light spots is easy to recognize in the clouds. The left side of Figure 9.2 looks marbled because the rising plumes experience condensation (cloud

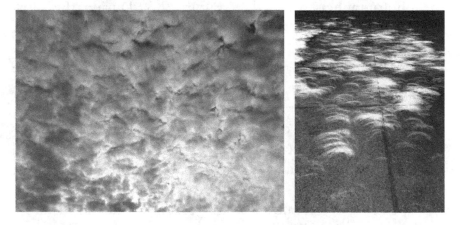

Figure 9.2 Dark and white spots in the tapestry of a cloud seen from below (left photograph), and how the spots may look when flying above the cloud (right photograph).

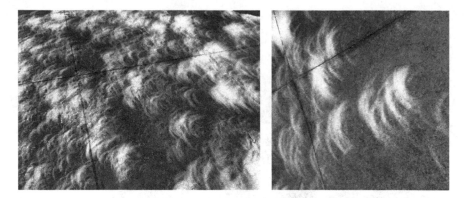

Figure 9.3   Crescents of light on the pavement during a partial eclipse of the sun. The photograph on the right side is a close-up of the area near the intersection of the gaps in the pavement.

formation), while the colder and drier air sinks into the spaces between the clouds. We have seen this phenomenon in Figure 7.1, where the clouds are shaped as bands (parallel rolls).

You might think that if you were flying above the clouds of Figure 9.2, the image would look like the photograph on the right side of the same figure. This is how the power to make connections can trick the mind into illusion, mirage, or worse (cf. Houdini quote, p. 6). The physics behind the right side of the photograph is revealed in Figure 9.3. During the most recent solar eclipse in North Carolina (August 21, 2017), the pavement behind our house was covered with a carpet of crescents of light. These had come from the sun, which at the time was shaped as a thin meniscus of light that reached the earth as a beam of parallel rays. Before reaching the pavement, the rays became a bundle of mini beams, each passing through its own "peephole" between a few tree leaves. The ground served as screen for the image of the light source, the solar meniscus.

Open eyes capture not only images that we all know (moon, sun) but also images that strike us as new, intriguing, and perhaps worthy of first-time exploration. Figure 9.4 is one

Figure 9.4   Dead flowers gather into equidistant bunches when blown by the wind against the curb.

example from the pandemic. Dead flowers fell from the trees all over the pavement, and the wind blew them against the curb. The wind did a lot more: it kneaded the flowers into equidistant bunches, each bunch the size of one shoe, and each spacing six times longer than that. Why such perfection? One reason may be that by being organized into bunches, the dead flowers pose less drag in the wind, which means that this way the wind sweeps the pavement more easily. The spacing between bunches may be due to a wind that blew perpendicularly against the curb, and in the quest for greater access, the wind constructed its own gates between bunches. Said another way, the bunches emerged in the spots where the wind stagnated (relatively) against the curb, before spilling over it. The location of the bunch over the stagnation area is akin to the gathering of the tea leaves in the center of the bottom of the cup, after stirring.

By the way, if you can open your eyes to the formation of moon craters, sunspots, eclipses, and clouds, then you can inject reality in the online experience of teachers and students. Advise both groups to stop, look, and listen. They will see images in front of them that even the best online teaching material cannot convey.

New experiences are the prescription for slowing the time and doing more with the time that you have. Time stopped during the terror attack of September 11. Change is the source of the feeling that you live longer. The saying "Time flies when you are having fun" seems to convey the opposite advice, but this impression is incorrect. The fun that does not last long enough is the feeling we enjoy when pleasant changes are taking place at high frequency and in great numbers. Weddings, movies with surprise endings, and upsets scored by our favorite football team are good examples. They seem too short because we are addicted to feeling good. Every event of this type lasts forever. We remember it, and we return to it time after time. These are the special events that slow down the mind time the most. And so, we arrived unwittingly at the oneness of pleasure and slow time, and why the two are good for human life.

Doing the same thing all the time is a killer. I was made aware of this danger as a pre-teen. Under communism in the '50s, the government (i.e., the communist party) required workers to compete among themselves to break records of productivity. The policy is known as Stakhanovism, after a celebrated coal-mining champion named Stakhanov, in the USSR. My mother was a pharmacist in the polyclinic of the state railroads, and she was bringing home horror stories about workers going crazy. One worker, a riveter, complained that, day and night, his mind saw and heard only one thing: "Bang a nit, bang a nit, bang a nit."

Such stories coincided with another, from inside our family. My father's brother was taken as a prisoner of war and forced to work in a coalmine for eight years. According to my parents, by the time he was released, he had gone insane. He recovered eventually, just like my mother's patients in the pharmacy. Stakhanovism was such a monstrous idea that the communist regimes abandoned it in the late '50s.

Prolonged isolation (prison, labor camp, lockdown) is damaging, yet the damage can be removed at least partially by returning to social life and behavior. From my teen years, I remember a favorite book (with original lithographs) about how the life of Robinson Crusoe changed after he met Friday. It will be the same for most of us when the pandemic ends, but our lives will be affected forever. We are getting used to being online, in partial isolation.

Do not stick to the same routine, morning, noon, and night. Set the dinner table, clear the table. It's not easy to break the routine because 'daily' life means routine. It takes thought and effort to get away from this.

During the routine, the mind wanders and lives in prerecorded images of events, not in the new events that your senses transmit. Make the effort to get out of bed by promising yourself to start your day with something different. If you are right-handed, do your daily chores with your left hand as often as possible. You will have to think about it every time you do it. If you are a basketball player, you become a much better player.

This is not the worn-out advice to "think outside the box." On the contrary, think inside the box, because the box is you. What you think, what you do, who you are — these are designs that are free to be changed. Work in the box. You are the boxer and your own corner man.

Do not fall victim to your own opinion, which is the predictable, the routine, and the meat grinder of fast time. This danger becomes acute during times of political polarization. Do as I do every day: read one publication known for its bias and another publication known for the opposite bias. No, do not add and divide by two. Just read, and from around every corner you will be surprised by something new. Smart or stupid, does not matter. If smart, admire and use; if stupid, laugh and forget. What matters is the life of experiencing something different.

The new thing may contradict your opinion. That is good when it happens. Never buy your own material no matter how good you think your opinion is. Never believe in your own joke.

Humor is effective when it contains a good dose of truth. This is why the wise enjoy good ideas wrapped in humor. When a math teacher asked a student to find a square root, the wise student replied: "Sir, it would be much easier to find a round root." The student was right. Imagine that you must pull a carrot out of the ground. The required force is proportional to the surface of contact between carrot and soil. If the cross section of the root has two shapes, square and round, one can show that the force required to pull the round root is 11 percent smaller than the force required to pull the square root.

This is only the beginning of the path of inquiry opened by the student's humor. Now, we know that the carrot is round for a different reason, not to be pulled out easily. The carrot is round in order to be strong enough to bend in all directions, as is required by the winds that push its canopy in all directions. This is also why all the pivot-shaped roots and their extensions above ground (the trunks) have round cross sections.

Strength under multidirectional loading is why the teeth of many animals have round cross sections. They are round not because they must fall out easily. We arrived at an unexpected connection between vegetation design and denture design. The unexpected connections (idea, time, place) are very good when they happen, because they slow down the mind time. They are memorable.

After this, the teacher can formulate a math problem of his own: What shape should be the cross section of the nail that fastens the horseshoe to the hoof? Square is the answer, because the force that the shoe exerts on the nail is a pulling force, and the purpose of the nail is the purpose of the shoe, which is to remain fastened.

Metaphors, like humor and analogies, empower the mind to connect the new impression with previous impressions that are correct and tested through experience. When we speak of "inspiration," we mean the intangible act of acquiring an idea from somebody else. The intangible is made palpable by the word *inspiration*, which means the physical act of inhaling. It is the same with the word *soul*, another intangible that got its name from the physical act of exhaling (*souffle*, breath in French, *sufflatus* in Latin, and *suflet* in Romanian). That's correct, because no respiration means no inspiration and no soul.

New experiences are accessible to those who have the freedom to change. Freedom is the name for the multitude of physical characteristics that allow changes to occur. This is a physical fact, not a political opinion. In human society today, freedom of thought and movement goes hand in hand with access to power (watts), space, and wealth [17]. This reality has led to the speculation that the rich can use wealth to trick the mind time into slowing down for them [4, 5]. True, for a rich person, a vacation in Maui may make the time to slow down for a while, but soon enough the change loses its novelty, time speeds up again, and the jetsetter can't wait to get back to the office.

You don't have to be rich to treat yourself to something new. You must work at it. Visit a nearby farm if you are a city person. Learn a Romance language if your native language originates from east of the Urals. You don't have to fly to Zanzibar for a new hotel and beach experience. Look in your immediate region, drive one or two hours, and treat yourself to more "slow time."

As we saw in Chapter 2, in an older person the number of new impressions is reduced, and the feeling that the day ended too soon is depressing. From this comes the tendency to make the day last longer. This can become a habit of staying up late, which is a mistake because it has the opposite of the intended outcome. The mistake leads to sleep deprivation, which shortens the lifespan. Lack of regular sleep is a killer.

The tendency to make the day last longer sneaks up on us unannounced. My former doctoral student Jong Lim [20], now a professor at Zhejiang University in China, wrote to me that the French geologist Michel Siffre entered a dark cave on 16 July 1962, where he planned to remain for two months. Tracking the days according to his sleep pattern (one night's sleep equals one day), he believed that his underground stay was ending on August 20. Instead, when he emerged it was 14 September, 25 days later. With little new to observe, his mind stayed awake every night waiting for more new things to happen before sleep.

We are surprised by the new and unexpected, we like it, and we remember it. Artists benefit from this truth, intentionally and unintentionally. Good art is original art, defined by something new. The Beatles revolutionized rock music with their "new sound." Never mind the American rock that influenced them, and the subsequent successful bands that were influenced by them. Elvis Presley's immortality is due to the same effect, and so is the immortality of the stars influenced by him, Johnny Hallyday in French rock and Adriano Celentano in Italian rock.

Immortality does not come from having a single success. The immortal is the opposite of the "one-hit wonder." Immortality follows those who after applause on stage always have something new for an *encore*! This is obvious in the creative careers mentioned above, and it is equally obvious in the creative enterprise of science. Ludwig Prandtl's fame does not rest on his first big idea, boundary layer theory (Chapter 8). It rests on his entire career, which is a pyramid of big blocks in the construction of modern fluid dynamics, laminar and turbulent.

The immortality of a plagiarist does not come from a single act of stealing credit. The one-hit wonder among plagiarists is easily forgotten; in fact, this is why the science enterprise turns a blind eye on such cases. Scientists who are truly creative do not have free time to police, or to serve as administrators whose job it is to enforce ethics. The truly creative console themselves with the thought that it is better to be stolen from than to have to steal.

Immortal plagiarists are those who plagiarize repeatedly. They are the professional plagiarists, the epigones. That's one way to become famous, and these are placed automatically in the camp of the "good because famous," which is not to be confused with the sparse huts of the "famous because good."

The "new" that the mind records is a new *connection* made between images. The new is not every new image. Nature speaks to us in the language of images that she has taught us already. The human mind has the natural urge to recognize and under-stand, which means to rationalize, to explain, and to simplify what it needs to retrieve, i.e., to remember more easily. It stores the imagined and the unseen in the imagery that nature has already taught us. This is where the observed image lands on the mental movie screen. The result is that similarity, dissimilarity, and analogy occur in the mind, and they are useful. This is the tendency that empowered humans with speech, writing, drawing, painting, and design of contrivances. What I just wrote is itself a connection, and here are two accidents that opened my eyes to it:

A few years ago, I was flying from Hong Kong back to the U.S. [18] and on the TV screen in front of everybody appeared the flight path and current progress (Figure 9.5). Tens of thou-sands of passengers must have seen this image before. Obviously, the image is China and Japan. I was struck by an entirely differ-ent and much more familiar image on the sea floor. It is a woman wearing Japan as a scarf. Her hand is Taiwan, and her purse is the Philippines. Her head and hairdo are the Sea of Japan. Hidden under the sea is what the school does not teach, but human evolution does. The bottom of the sea speaks to us

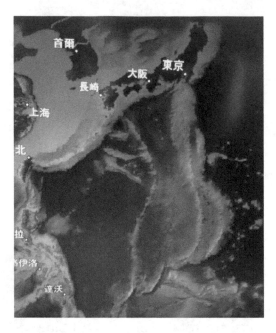

Figure 9.5 Nature speaks to us in the language of images that she taught us already: New mental connection while flying over Taiwan, and new connection while seeing a fallen leaf of red tip photinia.

in a prehistoric language that we all understand, with a message that we can articulate, remember, and transmit.

The second accident, also in Figure 9.5, took place during the pandemic as I was testing my theory of how to slow down the time by walking in the park. I saw a fallen leaf of red tip photinia, which is an evergreen shrub very common in our neighborhood. My first impression was the shape and the red color, which made me think of a woman's lips. Sure, I knew it was a dead leaf, but I had to go home to check the encyclopedia to learn the name of the plant. Perhaps, a first impression was what inspired Jacques Prévert to write the lyrics to the famous song "*Les feuilles mortes*" (The dead leaves, 1945). The song became popular in English as "Autumn leaves," which illustrates how much can be lost in a translation.

Proportions are everywhere, yet they surprise you if you keep the eyes of your mind open. If you question the proportions, they surprise you even more. This way they change your view of your surroundings, and the result is... two results: your perceived time slows down, and in the future, when you remember to question proportions, your time slows down again. Here are two examples from my experimenting with this idea during the lockdown: bird nests and city governments.

Big birds make big nests, and small birds make small nests. This holds true for bird species all over the globe, which is why two very different birds, the osprey from the North Carolina coast and the white-browed sparrow-weaver from the Kalahari Desert in southern Africa, are shown in Figure 9.6. Why does the proportionality happen?

There can be many explanations, and all rest on the innate urge of the animal to move (to live) more easily, i.e, more economically. The size of the organ emerges naturally as a trade-off (one choice, a compromise) between two competing extremes,

Figure 9.6 Big birds have big nests, small birds have small nests: osprey on the North Carolina coast (left), white-browed sparrow-weaver in the Kalahari Desert, and bald eagle on Mount Vernon, Virginia (Photo William Bejan, with permission).

both very bad for the living system. The "organ" in this example is the nest, which is known as "niche" in biology, and as immediate "environment" in physics and thermodynamics. The nest belongs to the bird, and the bird, eggs, and chicks belong to the nest.

Here is the clash between the two bad extremes:

First, the bird loses more body heat at night if the nest is small, i.e., a poor thermal insulator with negligible thickness. Losing more body heat requires finding more food to eat. The tendency then is to build a big nest, which constitutes a thicker air-layer insulation around the warm body.

Second, the bird must scrounge for tiny twigs and fibers over the immediate territory, and then it must carry this material to the nest site. Transportation requires work, dissipates the work, and deep down (in the bird's belly) requires food. In this second extreme, a thicker nest requires a wider search for food, more flying, and more food consumed.

Either way, the bird suffers. Finding food is a permanent ordeal for the hungry. The only option (option is the keyword in place of the so-called "optimization") is at the intersection between these two extremes. The result is that the nest size is neither too small nor too big, and so is the size of the "niche" from which the material is collected to build the nest, which is the most immediate portion of the niche that the bird touches. At the trade-off, the bird size and the nest size are comparable.

Here is a simpler vision of the physics behind the bird nest proportion. In cold climates, the fur coat is thinner on the bigger animal than on the smaller one. Compare the horse and tiger with the sheep and the wolf. Even better, think of the biggest warm-blooded animal you know. The thickness of the coat is the trade-off between the rate of body heat loss and the power spent on carrying the coat.

Optimization is a trade-off, one choice, the single act of *opting* [21] for the trade-off, the compromise, the natural selection. Optimization is not an infinite number of options (dots) that make a continuous curve that looks like a bucket or a haystack.

Large groups (cities, countries) have big governments, and small groups have small governments. This proportionality is true, broadly speaking, and the obvious proportionality is worth questioning, even if it is approximate.

If the government is open to being questioned and to being changed, then society arrives unnoticeably at the proper (persistent) balance between the size of the population, the number of its laws, and the size of its government. As in the example with the bird nests, the "size" appears naturally as a trade-off between the two extremes: too much government versus not enough government. At the intersection of the extremes, the government size that emerges is roughly proportional to the size of the population.

Here is an earlier manifestation of this phenomenon, from European peasant society. Every village has a characteristic number of dogs. A large village has more dogs than a small village. The small village also has fewer sheep, chickens, and houses to protect. The size of the "dog force" is roughly proportional to the size of the population that is being protected and that feeds the dogs. It is the same with any other institution and component of government.

If the dogs are too many, then everybody works to feed the dogs while being afraid of them. If the dogs are too few, then the sheep and the chickens are eaten by the wolves while the villagers are robbed, killed, and burned by the Huns and the Mongols. The village dies, and the few survivors flee. From the competition between two horrible extremes emerges the natural size, better known as the "not too big and not too small" size.

The questioning of natural proportionalities peels the cover off an even bigger idea. In the bird nest example, the proportionality is between animal size and the size of its natural niche. The nest is one example of niche. In the village example, the villagers form the niche that sustains the dogs. In the growth of cancerous cellular clusters (Figure 8.11), the extracellular matrix is the niche that nourishes the cells. In all such examples, the niche is the immediate environment of the object (or system) of interest.

The bigger idea is the oneness of system and environment. The attraction to the Impressionists in the late 1800s stemmed from this new view. They painted "whole" images and impressions, instead of the most realistic and fine details demanded by the prevailing doctrine in art schools.

Proportionalities reveal that the object is as important as its immediate environment. This direction of thought runs against reductionism. It opens the mind. This thought is timely because all of science is handicapped by the unquestioned focus on the system (the small, in the center) as distinct from its environment, which is usually dismissed. This handicap covers the board from physics (thermodynamics) to biology, where the even smaller system is seen as even more important. What is system today becomes the newly dismissed environment of the even smaller system tomorrow.

The bigger idea is that the power to predict nature and its future evolution comes from recognizing the whole, the object (the system) in its immediate surroundings (the niche). Greater power to predict is like the power that comes from knowing more than one language. Reductionism may still serve its purpose, but even better is to be able to march in both directions, along and against method, every time questioning the not mentioned direction of the march.

To take the holistic approach does not mean to march blindly toward megascales either. The whole (object and niche) may in fact be very small, if that is where the questioning leads the thinker. To take the holistic view is to free yourself from the rigidity of compartmentalized science, and to become capable of anticipating and putting on display what reductionism and Darwinism never could: the physics of form, configurations, images, drawings, evolving designs, and evolution.

Holism should not be confused with being shallow. On the contrary, holism means to possess the disciplines, the organs, the assembly, and the purpose of each. Holism calls for having more polymaths around. We need them because the best scientific breakthroughs come from free-minded individuals who excel at more than one thing. Holism is not to be confused with the interdisciplinary, which (if emphasized too early) translates into shallowness, i.e., not knowing the disciplines while claiming to know the tent that hovers over them.

To see the things that are worth connecting, one must have the time to contemplate. If the train travels at high speed, the images that sweep the window are a blur. The passengers do not even notice; instead, they read and type on what is on their lap. Like the train, the mind time races on.

The train may be going far, but the construction of connections in the mind does not. The reverse is also true because slow exploration builds tall and deep edifices in the mind. This contradiction is far from new, and if we open our eyes, we see it in many places. Its oldest version is present in Aesop's fable "The Tortoise and the Hare" [19].

The monotony of a quiet life stimulates the creative mind.

Albert Einstein

The finest creative thought comes not of organized teams but out of the quiet of one's own world.

Theodore von Kármán

The story of this book was told in images because that is how we all think, in mental viewings called "ideas": Connections take place between viewings that have features in common with older viewings, which have left impressions deeper than other connected images. My own introduction to this phenomenon happened during my first weeks in America. The English language was new to me. I was dreaming in my old world, in my hometown, and my maternal language. Two months later, I noticed in my dreams that events that had begun on the street where I grew up were continuing seamlessly on the street of my famous dorm at MIT, called Senior House. The street was an image that had impressed my mind an immense number of times, and now it was connecting my past with my present.

Viewed in perspective, all streets are tapered, like the two photographs mounted side by side in Figure 9.7. There are no parallel lines in the images. At first glance, just like in a dream,

Figure 9.7   A dreamer's train ride from the north of England to the south of France (Photo: Sylvie Lorente, with permission).

the lines transport the viewer by train from the north of England to the south of France. In our terrestrial lives this is a very difficult trip to make, yet in a dream it takes no time to happen. In a dream, tripping happens.

It may seem that at the fork in the road of life, the individual was forced to choose survival instead of dreaming. This book showed that dreaming and survival are one, and both are covered by the same physics principle as the perceptions of time and beauty.

People like to say that time flies like an arrow. I am not so sure. Like sand, time is slipping through our fingers and does not come back. This book is about how to tighten our fingers a little more.

References

1. A. Bejan, Why the days seem shorter as we get older, *Eur. Rev.*, **27**, March 2019, pp. 187–194, DOI: 10.1017/S1062798718000741.
2. Rachel Schnalzer, Is time flying by oddly quickly during COVID-19? Here's why you may feel that way, *Los Angeles Times*, 2 May 2020.
3. Amelia Diamond, Time seems to move weirdly right now. Here's Why, *Bernard & Hawkes*, 9 May 2020.
4. Doug Johnson, How you perceive time may depend on your income, *National Geographic*, 22 September 2020.
5. Stacy Liberatore, *Daily Mail*, People who can afford exciting experiences like vacations and hobbies produce more memories that are easier to recall, making them believe they have lived longer, study reveals, 22 September 2020.
6. Andrew Haffner, Running down the clock, *Southeast Asia Globe*, 3 April 2020.
7. Douglas Quan, The COVID-19 time warp: Why it feels like time has slowed during this pandemic, *The Toronto Star*, 4 April 2020.
8. Edward Browne, The surprising reason why time could be going quickly during lockdown, *Express*, 4 May 2020.
9. Christie Hawkes, Why does time speed up as we age? *So What? Now What?* 16 June 2020.

10. Luis Ignacio Brusco, La evaluación del tiempo en cuarentena, *BAE Negocios*, 17 June 2020.
11. Jessica Wagener, Darum vergeht die Zeit schneller, wenn wir älter werden, *ze.tt, ZEIT ONLINE*, 1 September 2020.
12. World News Platform, Frankfurter Rundschau, *Warum die Zeit schneller vergeht, wenn wir alt werden — und wie die innere Uhr wieder langsamer tickt*, 21 September 2020.
13. Marie-Ève Lambert, La journée, c'est long longtemps ..., La voix de l'est, *leSoleil*, 11 October 2020.
14. Maria Rattray, The hare and the tortoise: an allegory about life, *Medium*, 22 September 2020.
15. Carla Baum, Das unendliche Jetzt, *Die Zeit*, 20 November 2020.
16. A. Bejan, *Convection Heat Transfer*, 4th ed., Wiley, Hoboken, 2013, section 5.5.
17. A. Bejan, *Freedom and Evolution*, Springer Nature, New York, 2020.
18. A. Bejan, *The Physics of Life*, St. Martin's Press, New York, 2016, p. 65.
19. A. Bejan, U. Gunes, J. D. Charles and B. Sahin, The fastest animals are neither the biggest nor the fastest over lifetime, *Sci. Rep.*, **8**, 2018, 12925.
20. J. S. Lim, *Quality Management in Engineering*, CRC Press, Boca Raton, FL, 2020.
21. A. Bejan, Constructal law: optimization as design evolution, *J. Heat Transf.*, **137**, 2015, new paper: 061003.

10

Design science

The human events recounted in this story are like the spontaneous discoveries that add color and shade to the body of this book. They are simple acts of free movement and thinking, accessible when one is empowered by the view that evolution is physics, i.e., an image that is present in all objects with movement and freedom to change. They are about new mental viewings, drawings, changing images, and moving images, in short, the evolution of useful things called ideas.

They all fit into the rich tapestry of Science, Technology, Engineering, and Math, which some academics now refer to as "STEM." Wow, at the end of this story, I feel like Monsieur Jourdain from Molière's play *The Bourgeois Gentleman*: Well, what do you know about that! These forty years, I've been doing STEM without knowing it!

From time to beauty and design, and now STEM, what is the connection?

Before the connection, there were the coincidences in the observations. In my view of nature and the history of science, coincidences scream at some of us to discover the principle (the common cause) that hides behind them. After the discovery of the principle, the coincidences vanish, and so do their novelty and surprise. Science moves on, more powerful and easier flowing than ever before.

Every aha we hear on the path from coincidences to connections is a Gordian knot that gets cut in one blow, with a heavy axe. Coincidences happen, like the agreement between the falling brick and the gap that awaits it in the Tetris game. Once inside the gap, the brick and the wall are the new connection. The wall has several waiting gaps because its many bricks are the earlier connections, from earlier coincidences that screamed for a principle. The new principle is the new physics.

In short, human perceptions (fast time, beauty, etc.) and the urge to design and communicate share the same physics. They are about the evolving flow architectures inside the human body and outside, on earth, in society and beyond. They are all evolving into flow configurations that provide greater access to what flows.

Time is change. Time is as real as the shape and object. The reality of time is subjected to test and validation everywhere. From a barn on fire, horse, pig and man run for their lives: the perceived change is the same because it is real.

The new physics that you just learned is like some "free money" that you just saved in your bank account. It is like free gasoline at pumps everywhere. I reduced this mental viewing of the "everywhere" to just two ideas:

(i) First, what is the saved money, as physics? In my life the physics meaning of money was an unexpected human event, just like the three cases illustrated at the end of Chapter 8. A few years back, my colleagues and I discovered the physics that underpins the occurrence and evolution of economics and sustainability [1–5], which I summarized in a 2020 book [6]. Then, as if by chance, a Turk, a Brazilian, and an American walked into a bar. The bartender asked: What is money? Here is what came out of that conversation, which we just published, after the 2020 book [7]:

Money saved by an individual (or group) is a measure of the physical movement of what the individual could not execute on the spot because the power that he generated exceeded his ability to consume it, that is, to destroy it through his own movement. Excess power is unused movement, and it is stored for later use.

Money (power, movement) saved at location A is stored in a bank at A. Such savings are used later to sustain movement at A and in many other places. The invention of money and savings increased the spread of movement on the landscape as a "chain reaction," as illustrated in Figure 10.1.

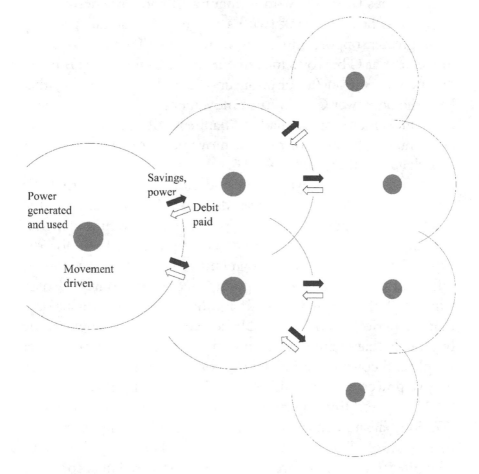

Figure 10.1 Movement spreads as a chain reaction on a territory when power savings (money) accompany the generation and use of power at discrete centers, which borrow power from and return (save) power to each other. The movement driven by each center covers a local area, around the center.

Saved power is the essential feature of the economic flow system because it endows the system with time-dependent aspects of behavior, which in physics are known as inertia, storage, and retrieval. One cannot bank the power generated by burning a fuel, but with money (capital) one can bank the movement that *will be* driven by that power, so that the movement stored by convention (as money) at A and invested later at B emerges later as new movement at B and elsewhere.

Debt is how the bank (with saved money, capital) provides even greater opportunities for movement. The generator of movement at C borrows movement stored as money at B based on the convention (a common understanding in society) that later the borrower C will return movement to B. This way, even more movement can be made to happen at A, B, and anywhere else. Indeed, because of money, movement grows and spreads as a chain reaction; cf. Figure 10.1.

Wise experts in government and academia lament that the power system of the U.S. was designed by disregarding the "fact" that the power needs, supply, and use are time dependent, not constant. This apparent error serves as basis for the experts' notoriety and employment. The experts are correct only half-way, because they point to obvious sources (solar, wind) and lifestyle (lighting, air conditioning, night versus day) that are periodic, not steady. Neglected in this argument is the big source that maintains the time-dependent behavior of power supply and consumption. The big source is in the banks, where motive power is stored and retrieved periodically as money.

Money and free enterprise (capital, savings, debt, change of decision) facilitate human movement to levels far higher than in their absence. This is why money and banking happened naturally *after* trading in nature. This change occurred in accordance with the constructal time arrow of evolution toward flow configurations that provide easier and greater access. Access is achieved by freely morphing the flow configuration, where removing obstacles does not mean simplifying the drawing, or cutting away parts of the drawing. Compare human movement before and after the invention of money. Compare human movement before and after industrialization (power generation, capital). The change was toward more movement and complexity (hierarchy), and the change happened by itself, naturally. Or consider the opposite example: the severe slowdown in human movement that occurred after the elimination of private property and free enterprise during communism. The way to easier flow and greater access is by rearranging (morphing) the path, as in the duo of erosion and deposit (sedimentation) that governs the evolution of river channels.

The major changes that occurred were natural "next steps" to the adoption of fire, boating, domesticated animals, and the wheel. The changes occurred in one unmistakable direction: toward greater movement on earth. Such changes do not go away, because in evolution what works is kept.

The periodic end of economic expansion (movement, wealth) emerges as a natural, physical feature of the spreading movement, which has access to power (money), freedom to morph, and power storage (savings) for future movement on even greater areas. The movement is driven by power generation, which is interspaced with power savings on the same area and during the same time period. The theory [1–7] was constructed systematically from the physics basis of economics concepts (money, savings, time, bubbles, crises) to a physics that accounts for the time-dependent spread of movement on an area. Previous work had shown that physics accounts for the proportionality between the annual wealth (GDP) of a population and the annual consumption of fuel to generate power for that population [1, 2]. The theory of economics "as physics" extends this view to the more realistic situation where every movement in society (wealth, fuel consumption) is time dependent.

With the principle, we predict the *future state* of the connected whole, namely, its configuration, movement (flow), and
morphing in time [8]. We can now fast-forward the evolution
of that object, which morphs in time and space with freedom.
In human activities of artifact creation (education, technology,
medicine, transportation, wealth, etc.), we become more powerful. We endow ourselves with objects that become better,
faster, and cheaper. We do this with increasing confidence that
our new creations are better and, as a consequence, will have
greater staying power.

Coincidences are the stuff of human perception, and connections are the stuff of the memory in the brain. Memory, like
the brain, has finite size. Principles are how the connections are
organized (compacted) in memory. All three (coincidences,
connections, and principles) belong to the observer, and so
does the observer's ever-expanding domain of observation, the
niche called "environment" and "nature."

(ii) Second, what is the new physics that you just learned?
It is the new principle that came from new connections made
between perceptions reported as observations by others, not
by you. The observations were confirmed by others. The coincidences were tested and validated by others, not by you.
Coincidences were also noted by others, perhaps much earlier,
in antiquity. To discover the unpublished wisdom of the
ancient is the stuff of historians, and it is facilitated immensely
by the technologies of enhancing human perceptions (digging,
magnifying, deciphering) that new science makes available
daily. The principles discovered from verified coincidences
were and continue to be tested and verified by others, not by
you.

In this you see the subtle and beneficial effect of the *scientific method*. Ideas, results, and devices created by others have
already been *tested* (submitted to peer review and possible falsi-
fication). They are now placed in storage and made available to

everyone. The new user does not have to face the impossible task of coming up with the idea, testing it, and then proving it in a prototype and journal publication. All that work has been done for us by individuals, the altruism of whom goes unrecognized even though what they would like to receive in return is very little: to be acknowledged as creators (cf. Chapter 6). Their unending contributions are akin to philanthropy, about which James Feldstein wrote [9]: "The giving of money represents motion on the part of the donor; the amount of the gift represents the magnitude of that motion. The event of a gift results from an action or series of actions."

That is the new physics, and it is being deposited in a free bank by many observers, a few questioners of coincidences, and even fewer seers of the much fewer principles. Not by you.

You, like me, are the beneficiaries, the lucky to be living in this unparalleled era of advancement. The free bank is called physics, or science, more broadly. All you must do is enter the bank, although access to it is not uniformly plentiful on the globe, as it requires parents, education, a culture of questioning, freedom, and belief in the individual (cf. p. 24).

The reproducibility of the scientific ideas and results validated by others and contributed to the free bank of physics is like the accountability and solvency of your real bank. You use both banks (science and real bank) with confidence, and you become a more powerful human and machine specimen. You come to the realization that people have been changing the life of the earth forever: the animals, the climate, the lakes, and the streams. The human impact became particularly acute after the taming of fire.

> From knowledge comes foresight, from foresight comes action.
> (*Science d'où prévoyance, prévoyance d'où action.*)
>
> Auguste Comte

Perceptions and good art trigger emotions, and emotions trigger action. The word *emotion* comes from the verb *émouvoir* in French (to stir up), the origin of which is the Latin verb *emovere* (*e*, out + *movere*, to move, a visible reaction). Good art triggers more impressions (feelings, urges), in addition to the mind time and beauty covered in this book. Examples include comparative thinking (the urge to have more), envy, fear, hesitation, lack of confidence, mourning, and so on.

The point is that with the new idea of physics, you are empowered to a level higher than before. Empowered in every respect (movement, wealth, health, longevity). Every step of innovation is measurable as an increase in power (watts) produced, consumed, and saved to be lent to others to produce and consume (cf. Figure 10.1). For this reason, the new physics entrains and empowers the whole world in a new chain reaction (Figure 10.2).

The trigger of this new kind of nuclear explosion is known to those who studied the history of the science that is practiced today. The physics of economics (Figure 10.1) is just one filter through which we can grasp the much more general message of Figure 10.2. If you have heard the saying "Have money, will travel," you can join the three jokers in the drinking bar and teach them and the bartender a new saying, "Have physics, will travel."

Get to know history and the physics of evolution. If you practice how to predict the past, you will discover that you are able to predict future changes. You may even figure out what changes to make so that future generations do not suffer the horrors of the past. Academia needs more polymaths, to elevate students with history and art in addition to science. The polymaths come from all walks of life [10–19] and deserve to be read and heard.

The oneness of nature, design, and engineering stimulated Leonardo da Vinci, and even though he illustrated it clearly and

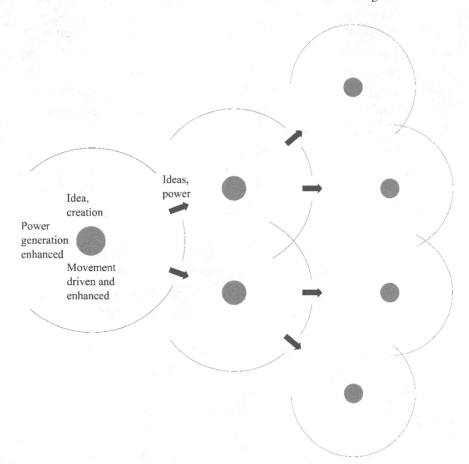

Figure 10.2   Ideas flow from high to low: from those who create them, to those who need and use them. Over the entire population, one innovation empowers and facilitates everyone's movement (life).

beautifully, people are still mesmerized by narratives of the disunity (diversity, randomness) and the invisible that comes from reductionism. For them, it is fashionable to read that "you cannot predict the future." I am not so sure about that. It is evident to me that science is about enhancing the human ability to anticipate the future, everything, from the weather to avalanches, human behavior, and the reach of the pandemic.

STEM is an important gate of access to the science of design. Engineering contributes the reliable truths of the physics of power: everything moves because it is pushed, and the movement destroys the power used. Engineering contributes the holistic view: objects do not exist and persist in isolation. They have power, movement, directionality, purpose, evolution, user, niche, and sustainability. Engineering contributes the physics of freedom, which is better known as design and evolution in nature. This is the freedom to create, to change, and to be forward-looking and not afraid of proposing the unthinkable.

Resilience is an engineering concept. An object has resilience when it has the ability to bounce back, to recover its shape and strength. In the human realm, this is a need every time and everywhere. The question is, what is the solution? How is one to improve or safeguard resilience?

The solution that I know is what I learned from basketball. To have resilience is to constantly make the effort to not fall out of shape, so that you have nothing to regain. That means you should never interrupt the training, because if you do, it is much more difficult to return to being a meaningful player.

During the pandemic, all of us were challenged to stay in the game. Some did it better, others not as well. The pandemic is still an opportunity to practice the solution method that I sketched above. In fact, if you did not practice it already, do it now because the pandemic is still here, and future pandemics are around the corner.

About the tendency to overlook engineering, Rankine wrote in 1859:

> The improvers of the mechanical arts were neglected by biographers and historians, from a mistaken prejudice against practice, as being inferior in dignity to contemplation; and even in the case of men such as Archytas [an ancient Greek philosopher] and Archimedes, who combined practical skill

with scientific knowledge, the records of their labours that have reached our time give but vague and imperfect accounts of their mechanical inventions, which are treated as matters of trifling importance in comparison with their philosophical speculations. The same prejudice, prevailing with increased strength during the Middle Ages, and aided by the prevalence of the belief in sorcery, rendered the records of the progress of practical mechanics, until the end of the fifteenth century, almost a blank. Those remarks apply, with peculiar force, to the history of those machines called PRIME MOVERS. [20, p. xv]

If you know the principles, you know where the design came from. That is where "reverse engineering" works: in the minds of those who know the science of design, the principle. Reverse engineering is the mental process where one uses reasoning to deduce how an artifact was made (weapon, software, airplane; cf. Figure 6.12). Artifacts are objects made by humans, and they empower the human and machine species (recall that the word *machine* means artifact, contrivance in general).

Many more objects happen by themselves, naturally (animals, rivers, winds, valuable minerals, and celestial bodies). The many are not artifacts. To claim that a person can "reverse-engineer nature" is incorrect. Objects that are not made by people are not contrivances. Nature was not made by a person. If you doubt that, consider the fact that nature on earth and elsewhere is much older than the biosphere of the earth. Rivers and winds like ours are visible on planets without biospheres.

One can examine, describe, and catalog objects that came about by themselves, and that is empiricism. On the other hand, if one questions why many of such objects look similar, or why they behave similarly, the idea that pops up in the mind of the curious is that coincidences are due to a common *cause.* This idea is theory, the new physics that empowers future observers not to waste time searching for the new idea.

Theory must not be confused with empiricism. They are eminently different, like male and female. For this very reason, they are useful when practiced together.

If you know the principle, you can not only predict consequences that are observable but also make unintended predictions. If you know how your home water system works, then when you open the faucet for hot water, you predict (you are confident) that hot water will arrive twenty seconds later. If you know fluid mechanics really well, your unintended prediction is that in the next room, the flow rate of hot water will decrease.

I've been making unexpected predictions throughout my career, even before I called them the "constructal law" in 1996. In retrospect, the pre-1996 ideas constitute a body of constructal *prefigurations*. The Gordian knot of turbulence is its own list of examples of this kind (p. 137). Even more numerous are the charts of the design space in which the thermodynamic performance of designs (power, refrigeration) increases from decade to decade, while still residing below the ceiling commanded by the second law. That ceiling is well understood now, and it has nothing to do with form (design) and evolution in nature.

The physics of evolution is in the gap between the designs possible at the time and the Carnot ceiling: the gap has the tendency to decrease over time. In my 1988 thermodynamics textbook [21], the evolution of the possible designs is presented in Figures 2.1, 8.1, 10.24, and 10.27, and in Table 1.2., I described the evolution of these flow designs in terms of tendencies toward economies of scale, complexity, efficiency, etc., which today fit under the constructal tent [6, 21–24].

Why are prefigurations important? They are because they illustrate the human origin, life, and evolution of the idea. Nature is the kingdom of configurations in space and time, which are accessible to our senses and instruments. Not all the imagined configurations are possible. The constructal law is about predicting (1) the invisible frontier between *the possible and impossible* flow architectures, and (2) how to push this frontier into the domain of the impossible, namely, with flow configurations that evolve with freedom. All animal and vegetation designs are so "mature" in their evolution that they reside on that frontier. This is also why they can now be predicted by invoking the principle of design evolution as physics (e.g., [23–26]).

The designs that are most valuable for technology evolution and the permanence of superior performance are the flow architectures that fall on (or close to) the frontier between the possible and the impossible. For example, in 2004 we used [6, 27] the flow through a duct to illustrate the design space of possible and impossible designs (Figure 10.3). This idea appeared in

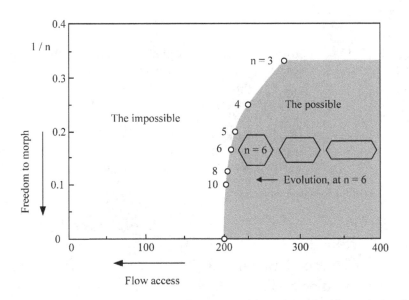

Figure 10.3 The possible and the impossible: How the flow through a duct finds easier access by means of greater freedom in morphing its design [6]. This figure was drawn after Figure 2 from Ref. 27. The measure on the abscissa is proportional to the overall flow resistance of the duct in laminar flow. On the ordinate, n is the number of sides of the polygonal shape of the duct's cross section. The points on the frontier between the possible and the impossible correspond to ducts with cross sections shaped as regular polygons. Three cross sections with $n = 6$ are aligned horizontally to the right of the frontier.

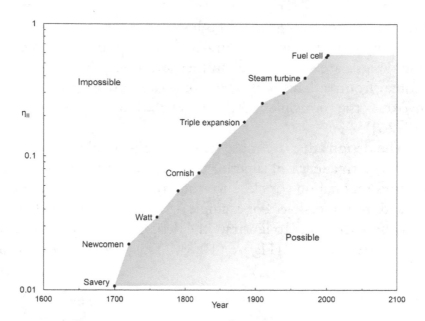

Figure 10.4 The possible and the impossible: highlights in the evolution of the second-law efficiencies of steam power plants during their history. The second-law efficiency η_{II} of a power plant is the actual efficiency divided by the efficiency of the same plant in the ideal (reversible, Carnot) limit. The data that mark the crest of the domain of possible designs are from Ref. 6. Note the logarithmic scale on the ordinate, and the $\eta_{II} = 1$ ceiling that no design can surpass. At every point in time, the crest divides the designs into the possible and the impossible; cf. Figures 2.1, 8.1, 10.24, and 10.27 and Table 1.2 from the 1988 edition of Ref. 21.

subsequent constructal articles and books of physics. Figure 10.4 shows a recent summary of old designs.

The science of design unites all the disciplines and specialties found in academia. The oneness of art and science can serve a university at all levels, not only in student curricula and professor-to-professor relations. It can do wonders if implemented at the highest level of university administration, starting with the president and the provost. Since the campus protests of 1969–1970, it seems that all the high posts in the top universities are filled by professors from the liberal arts. Uniformity of technocrats in the high posts would be just as bad.

Balance, rotation at the top, and open dialogue between administrators with antipodal backgrounds is the best medicine for an organization that does not flow and morph well. The same holds for all organizations, especially government.

Balance is everywhere in nature, as in the harmony between few large and many small, and few fast and many slow. Observed balance is an opportunity to review the main ideas that we shared in this book, from scanning and beauty to contrast, perspective and art. Figure 10.5 is a beautiful illustration of the balance perceived by the viewer. The cathedral is in the center, and the world around it reaches to us in perspective.

The painting has balance, all right, but the cathedral is not in the center of it. The cathedral is to the right of the vertical midline of the frame (go ahead, measure it). The balance perceived in the painting is because the left half of the image is rich in detail, while the right half is relatively empty. The eyes are doing more scanning on the left, and, because of the cathedral,

Figure 10.5 Paris street scene with the Notre-Dame Cathedral (Photo of a copy of the painting by Antoine Blanchard).

when they jump from the left side to the right side, they convey the impression that they crossed the midline of the image. The balance between scanning (with saccades) the two sides of the image has the effect of placing the "mental cathedral" in the center of the image. The balance was created by its famous painter, Antoine Blanchard (real name Marcel Masson, 1910–1988).

The evolution of the ideas that led to this book is a perfect illustration of the effect of evolution in human thought. My changing views took discernible shape in the books *Advanced Engineering Thermodynamics* (1997) [21], *Shape and Structure: From Engineering to Nature* (2000) [22], and *Design in Nature* (2012) [23], which centered on the law of the physics of design and evolution: the constructal law (p. 2).

The Physics of Life (2016) [24] closed in on two aspects of design in nature that belong in physics — life and evolution — even though they are known widely as pivotal concepts in a special domain, biology. *Freedom and Evolution* (2020) [6] brought the reader even closer to what is most important to him or her personally. What is important is the urge to know, to improve life, and to predict what will happen tomorrow.

Time and Beauty adds human perceptions to the physics of evolution, from mind time and beauty to contrast, shape, perspective, idea, and the oneness of art, design and science. Constructal law rules all of art and science viewings (ideas) and the human urges to have more art and more science.

I wrote this book during the coronavirus pandemic of 2019–2021. It was during this murderous period that the new physics revealed its usefulness. This is why I end the story with a brief summary, for the benefit of future generations.

Your life, every hour and every day, is a sequence of changes that you make — right versus wrong, up versus down, and so on, endlessly. You make decisions every moment you perceive a click. That's your freedom. Your life is the design, and you are the designer. You can choose to live one way or another, but you have to think about the change. Even the decision to think about every decision is a change, because previously you were not thinking about it.

So, it's about you. Making changes that are not routine will be good for your life. Luck in life comes from freedom to have more choices, more turns at throwing the dice.

References

1. A. Bejan, Why we want power: economics is physics, *Int. J. Heat Mass Transf.*, **55**, 2012, pp. 4929–4935.
2. A. Bejan, Sustainability: the water and energy problem, and the natural design solution, *Eur. Rev.*, **23**(04), 2015, pp. 481–488, Doi:10.1017/S1062798715000216.
3. A. Bejan and S. Lorente, The constructal law origin of the logistics S curve, *J. Appl. Phys.*, **110**(2), 2011, 024901, DOI:10.1063/1.3606555.
4. A. Bejan and M. R. Errera, Wealth inequality: the physics basis, *J. Appl. Phys.*, **121**(12), 2017, 124903, DOI:10.1063/1.4977962.
5. A. Bejan, A. Almerbati and S. Lorente, Economies of scale: The physics basis, *J. Appl. Phys.*, **121**(4), 2017, 044907, DOI:10.1063/1.4974962.
6. A. Bejan, *Freedom and Evolution: Hierarchy in Nature, Society and Science*, Springer Nature, New York, 2020.
7. A. Bejan, M. R. Errera, and U. Gunes, Energy theory of periodic economic growth, *Int. J. Energy Res.*, **44**, 2020, pp. 5231–5242.
8. A. Bejan and S. Gucluer, Morphing the design to go with the times, *Int. Commun. Heat Mass Transf.*, **120**, 2021, 104837.
9. J. Feldstein, Intellectual Capital, *Feldsteinco*, www.Feldsteinco.com.
10. Po Chung, *Designed to Win*, Leaders Press, Lightning Source, UK, 2019.
11. Madan Birla, *FedEx Delivers*, Wiley, Hoboken, NJ, 2005.
12. Madan Birla, *Unleashing Creativity*, Wiley, Hoboken, NJ, 2014.
13. Max Borders, *After Collapse*, Social Evolution, 2021.
14. Jarl Jensen, *The Big Solution: Deactivating the Ticking Bomb of Today's Economy*, Forbes Books, Charleston, SC, 2021.
15. Nikos Acuña, *Mindshare: Igniting Creativity and Innovation Through Design Intelligence*, Motion, Henderson, NV, 2012.

16. M. T. Takac, *Scientific Proof of Our Unalienable Rights*, 5th ed., Robertson Publishing, 2019.

17. Parag Khana, *Move*, Scribner, New York, 2021.

18. Fanis Grammenos and G.R. Lovegrove, *Remaking the City Grid*, McFarland & Co., Jefferson NC, 2015.

19. Sean Adams, *How Design Makes Us Think*, Princeton Architectural Press, New York, 2021.

20. W. J. M. Rankine, *A Manual of the Steam Engine and Other Prime Movers*, 12th ed., revised by W. J. Millar, Charles Griffin & Co., London, 1888.

21. A. Bejan, *Advanced Engineering Thermodynamics*, 2nd ed., Wiley, New York, 1997; 1st edition 1988.

22. A. Bejan, *Shape and Structure: From Engineering to Nature*, Cambridge University Press, Cambridge UK, 2000.

23. A. Bejan and J. P. Zane, *Design in Nature*, Doubleday, New York, 2012.

24. A. Bejan, *The Physics of Life: The Evolution of Everything*, St. Martin's Press, New York, 2016.

25. A. Bejan and J. H. Marden, Unifying constructal theory for scale effects in running, swimming and flying, *J. Exp. Biol.*, **209**, 2006, pp. 238–248.

26. A. Bejan, S. Lorente, and J. Lee, Unifying constructal theory of tree roots, canopies and forests, *J. Theor. Biol.*, **254**(3), 7 October 2008, pp. 529–540.

27. A. Bejan and S. Lorente, The constructal law and the thermodynamics of flow systems with configuration, *Int. J. Heat Mass Transf.*, **47**, 2004, pp. 3203–3214.

Index

CPSIA information can be obtained
at www.ICGtesting.com
Printed in the USA
JSHW032046170322
23717JS00001BA/32